Engineering Guide to Planning and Design of Virtual Power Plant

Feng Yao

Engineering Guide to Planning and Design of Virtual Power Plant

Hand to Hand Guide from Starter to Professional

Springer

Feng Yao
Department of Engineering and Research
China Huadian Engineering Corporation
Beijing, China

ISBN 978-981-95-2230-9 ISBN 978-981-95-2231-6 (eBook)
https://doi.org/10.1007/978-981-95-2231-6

This Springer imprint is published by the registered company Springer Nature Singapore Pte Ltd.
The registered company address is: 152 Beach Road, #21-01/04 Gateway East, Singapore 189721, Singapore

To my dearest family

This book, born from the intricate interplay of technology and engineering experiences, While, without your unwavering kindness, quiet devotion, and steadfast support, nothing could have been possible.

To you, from the depths of my heart, who made every word meaningful.

Preface

Engineering guidebooks typically cater to professionals and researchers, assuming a baseline of technical expertise. This book, however, endeavors to bridge the knowledge gap—speaking to amateurs, practitioners, and experts alike by balancing depth with accessibility. The goal is to demystify virtual power plants (VPPs) for all who engage with them, from curious newcomers to seasoned engineers.

Virtual power plants differ fundamentally from conventional power systems in their reliance on diverse stakeholders. Unlike traditional plants, where operations are centralized and controlled by a limited set of professionals, VPPs thrive on collaboration: homeowners with solar panels, businesses participating in demand response, grid operators, and energy service providers all play critical roles. Many of these participants—such as residential consumers adjusting their electricity use—may have little formal training in power systems, yet their involvement is key to a VPP's success. This book aims to equip every stakeholder, regardless of background, with the knowledge to contribute meaningfully.

Complicating this landscape is the integration of cutting-edge technologies: artificial intelligence optimizes energy, 5G enables real-time communication, and digital twins simulate complex operations. Translating these advanced concepts into understandable language without oversimplification is no small task, but it is essential. As VPPs emerge as a solution to pressing challenges—from managing the surge in electric vehicle demand to mitigating renewable energy fluctuations during typhoons—clear, inclusive guidance becomes urgent.

This book is our response to that call. It navigates technical details with rigor while ensuring clarity for non-specialists, connects theory to real-world applications, and highlights the roles of every participant. Whether you are a homeowner curious about solar integration, a technician deploying VPP systems, or a researcher advancing the field, we hope these pages serve as a reliable companion—illuminating the potential of virtual power plants to reshape our energy future.

Beijing, China
June 2025

Feng Yao

Acknowledgements The research presented in this document was primarily conducted at the **China Huadian Engineering Industrial Park** (CHEC Industrial Park), Beijing, China.

This work was supported by:

- China Huadian Engineering Corporation's Research Program (Grant No. CHECKJ22-01-32)

The author wishes to express the sincere gratitude to:

- Eng. Liu Xiaoping, Chief Engineer of Energy System at CHEC Industrial Park, for his guidance on market participation mechanisms
- The Industrial Park's Technical Committee for approving experimental protocols

Portions of this research were presented at the CEPSI 2023 Conference and received valuable feedback from the smart grid research community.

The author have no conflicts of interest to declare that are relevant to the content of this chapter.

All conclusions and opinions expressed remain the sole responsibility of the authors.

Beijing, China
September 2025

Competing Interests The author has no competing interests to declare that are relevant to the content of this manuscript.

Contents

Acronyms

5G	Fifth-generation wireless technology enabling low-latency, high-bandwidth communication for real-time VPP control and data transmission
AI	Artificial Intelligence, Technology integrating machine learning and predictive analytics to optimize VPP operations, including load forecasting, resource scheduling, and real-time anomaly detection
API	Application Programming Interface, Interfaces enabling software components to communicate, essential for VPP platform integration with external systems (e.g., power markets)
B/S	Browser/Server Structure, A software architecture where the VPP platform is accessed via browsers, enabling centralized management and remote access
BIM	Building Information Modeling, A 3D modeling technology used in VPPs for building energy management and optimization
DDE	Dynamic Data Exchange, A protocol for real-time data exchange between applications, used in VPPs for inter-system communication
DDG	Domestic Distributed Generation, Small-scale distributed power sources (e.g., residential solar) primarily serving local loads, with surplus energy fed to the grid
DERs	Distributed Energy Resources, Small-scale energy systems (e.g., solar panels, wind turbines, energy storage) aggregated by VPPs to form a unified, controllable power unit
DG	Distributed Generation, Power generation from small-scale resources (e.g., solar, wind) connected to low/medium-voltage grids, often integrated into VPPs
DSM	Demand Side Management, Strategies to optimize electricity consumption (e.g., time-of-use pricing) to balance grid demand and reduce peak loads
DSO	Distributed System Operator, A role within VPPs responsible for managing and operating the distributed energy system

DSR	Demand Side Response, User adjustments to electricity consumption in response to price signals or grid needs, coordinated by VPPs for grid stability
DSS	Distributed System Simulator, A software tool used for VPP load modeling, analysis, and optimization
ESC	Energy Storage Center, Facilities managing energy storage systems, often coordinated by VPPs for grid support
FM	Frequency Modulation, An ancillary service where VPPs adjust power output to maintain grid frequency stability, earning compensation
IBMS	Intelligent Building Management System, Systems integrating building operations, which VPPs can leverage for load control and energy efficiency
LLMs	Large Language Models, AI models (e.g., GPT-4) used in VPPs for natural language processing, automated reporting, and policy interpretation
LSTM	Long Short-Term Memory, A type of recurrent neural network used in VPPs for predictive maintenance and fault forecasting
MARL	Multi-Agent Reinforcement Learning, A technique enabling autonomous VPP management through coordinated AI agents, optimizing resource dispatch
MES	Manufacturing Execution System, Software integrating production scheduling and VPP load management to align energy use with manufacturing needs
NLP	Natural Language Processing, Technology allowing human-AI interaction in VPPs, such as querying energy data via chatbots
ODBC	Open Database Connectivity, A standard API for accessing database systems, supporting data integration in VPP management platforms
OPC	OLE for Process Control, A standard interface for integrating industrial automation systems, enabling data exchange between VPP components
PDG	Public Distributed Generation, Larger distributed power systems (e.g., wind farms) operated to sell electricity to the grid, often via renewable energy sources
PLC/HPLC	Power Line Communication/High-Power Line Communication, Technologies using power lines for data transmission, suitable for low-cost VPP communication in distributed settings
PPA	Power Policy Agent, An AI component within VPPs responsible for optimizing power allocation, load balancing, and frequency regulation
PUE	Power Usage Effectiveness, A key energy efficiency metric for data centers, monitored in VPP-integrated facilities to optimize power consumption

TSO	Transmission System Operator, Entities responsible for managing power transmission grids, to which VPPs provide ancillary services like frequency regulation
V2G	Vehicle-to-Grid, Technology allowing electric vehicles to feed power back to the grid, integrated into VPPs as mobile energy storage
VPP	Virtual Power Plant, A digital aggregation of DERs, controllable loads, and energy storage, enabling participation in grid scheduling and power market transactions

Part I
Starter

Get to know *Virtual Power Plant* as quick start.

Chapter 1
General Introduction

Abstract Virtual power plants are defined as aggregations of distributed resources using advanced tech to participate in power market and grid operations. Their characteristics include being digital, having diverse resources like controllable loads and energy storage, not needing large hardware investment, being a market participant, and differing from microgrids. Compared to traditional plants, they have diverse, eco-friendly resources and are competitive. The components are distributed power, distributed storage, controllable loads, etc., and they have various business and trading modes, with cross-regional trading being a key feature.

1.1 Get to Know About Virtual Power Plant from Starter to Professional

The book aims to provide a comprehensive understanding of virtual power plants, encompassing their characteristics, technical aspects, and operational considerations. Primarily, it targets technical professionals and researchers interested in virtual power plants. By delving into the technical details and practical applications, readers can gain a deeper understanding of the subject and apply the knowledge to their own projects. However, the book's extensive coverage ensures that a broader spectrum of interested parties can find relevant content tailored to their needs.

There are two key rationales behind this comprehensive approach.

- Firstly, virtual power plants aggregate resources from a diverse array of customers, involving numerous participants. Consider controllable loads in demand side response, for instance. Every ordinary power user has the opportunity to participate in power system dispatch by actively regulating their load, thereby reaping demand side response benefits. Through understanding virtual power plants, they can gain insights from this book on how to achieve these benefits instead of participating direct power generation. This knowledge is useful for those who wish to leverage the benefits of virtual power plants.

F. Yao, *Engineering Guide to Planning and Design of Virtual Power Plant*,
https://doi.org/10.1007/978-981-95-2231-6_1

- Secondly, virtual power plants are highly technology-intensive, incorporating a vast array of advanced technologies, including digital-twins, artificial intelligence, High-end communication etc. Grasping how to apply these technologies in practical scenarios poses a significant challenge.

This book intends to cover as much as materials and guide as many as readers step by step from Starter to professionals as of the following chapters.

Chapter 1: General Introduction

This chapter introduces the concept of the virtual power plant and starts from the fundamental ackowledgement on virtual power plant. Afterward, readers can search for information in corresponding chapter for the topics they are interested in.

Chapter 2: Comprehensive Conceptual Introduction to Virtual Power Plants

This chapter offers a thorough and in-depth conceptual introduction to virtual power plants.

From a technical point, it dissects the fundamental principles, key components, and operational mechanisms that underpin virtual power plants. It explores the various types of distributed energy resources that can be integrated into a virtual power plant. Additionally, it introduce the concepts behind, such as the communication and control technologies required to manage these diverse resources effectively, including smart grid technologies, real-time monitoring systems, and advanced control algorithms.

From a market transaction perspective, the economic models and market mechanisms associated with virtual power plants are introduced. It examines how virtual power plants participate in electricity markets, including wholesale and retail markets, and how they interact with other market players, such as power generators, electricity suppliers, and consumers. This chapter also explores the various revenue streams available to virtual power plants, such as energy trading, demand response, and ancillary services, and provides insights into the regulatory frameworks and policies that govern these market activities.

Chapter 3: Control Architecture of Virtual Power Plants

In this chapter, the focus is on the Control architecture and configuration of virtual power plants. It begins by presenting a high-level overview of the virtual power plant architecture, highlighting the different control architecture that make up the system.

Virtual power plant as an independent entity to participate in the power market transactions and grid scheduling, therefore, the virtual power plant within the various resources, need to be used to with the appropriate Control Architecture, flexible collaboration, scheduling and control. According to the different control structure of virtual power plant information flow transmission, the control mode of virtual power plant can be divided into: centralized control mode, decentralized control mode, and completely decentralized control mode.

Chapter 4: General Part of the Engineering Guide

This includes factors such as the size and capacity of the plant, the type of distributed energy resources to be integrated, the communication and control technologies to be used, and the economic and operational considerations.

Chapter 4 kicks off the engineering design guide, commencing with the essential considerations and solutions that need to be generalized. This chapter provides a comprehensive engineering design guide for virtual power plants. It begins by outlining the key considerations and requirements for the design of a virtual power plant, taking into account factors such as system reliability, efficiency, scalability, and cost-effectiveness. This chapter then proceeds to discuss the different stages of the engineering design process, from the initial concept development and feasibility study to the detailed design and implementation planning.

This chapter then provides detailed technical specifications and guidelines for each component of the virtual power plant. This includes the specifications for the physical layer, the communication layer, and the control layer. It also includes recommendations for the selection of appropriate hardware and software components, as well as the optimization of the system for maximum efficiency and performance.

Throughout the chapter, practical design guidelines and best practices are provided, covering topics such as system sizing, component selection, and integration with existing power systems. It also addresses common engineering challenges and issues that may arise during the design process, such as power quality management, grid connection requirements, and cybersecurity concerns, and offers practical solutions and mitigation strategies. By the end of this chapter, readers will have a comprehensive understanding of the engineering design process for virtual power plants and be equipped with the knowledge and tools necessary to undertake their own design projects.

Finally, this chapter provides a summary of the key design considerations and recommendations, and offers suggestions for future improvements and extensions to the virtual power plant architecture.

Chapter 5: Prerequisite Condition and Feasibility Determination

This chapter is dedicated to conducting in-depth research on the external conditions and determining the feasibility of virtual power plant projects. It begins by analyzing the regulatory environment, including the relevant laws, regulations, and policies that govern the development and operation of virtual power plants. This chapter examines how these regulations impact the project's viability, profitability, and compliance requirements, and provides insights into how to navigate the regulatory landscape effectively.

In addition to the regulatory environment, Chap. 5 also explores other external factors that can influence the feasibility of virtual power plant projects, such as market demand, energy prices, grid infrastructure, and environmental considerations. It presents a market analysis to assess the potential demand for virtual power plant services, including energy trading, demand response, and ancillary services, and evaluates the market competitiveness and potential revenue streams. This chapter also examines the existing grid infrastructure and its capacity to accommodate the integration of virtual power plants, and identifies any technical or infrastructure limitations that need to be addressed.

Based on the findings of the external conditions research, Chap. 5 presents a roadmap for comprehensive feasibility assessment framework for virtual power plant projects. It outlines the key indicators and metrics that should be used to evaluate the project's feasibility, such as economic viability, technical feasibility, environmental sustainability, and social acceptability, and provides a step-by-step approach for conducting the feasibility assessment. By the end of this chapter, readers will have a clear understanding of the factors that need to be considered when determining the feasibility of virtual power plant projects and be able to make decisions about the project's viability.

Chapter 6: Planning

Chapter 6 marks the entry into the planning phase, where the hardware and software market architecture and the path to realization are determined based on the conclusions drawn in the previous chapter.

It begins by outline the market-viable hardware and software components required for virtual power plant projects, including energy management systems, control centers, communication devices, and software applications.

Then proceeds to develop a comprehensive hardware and software architecture for the virtual power plant project. It defines the functional requirements, interfaces, and integration requirements for each component, and outlines the overall system design and configuration. The chapter also provides a guide to implementation plan, including the project timeline, milestones, resource requirements, and budget, and identifies the key risks and challenges associated with the implementation process and the corresponding mitigation strategies.

In addition to the hardware and software architecture, Chap. 6 also addresses the organizational and management aspects of the virtual power plant project. It defines the roles and responsibilities of the different stakeholders, including the project owner, operator, suppliers, and contractors, and outlines the project governance structure and decision-making processes. The chapter also provides guidance on how to manage the project's risks, costs, schedules, and quality, and how to ensure effective communication and collaboration among the different stakeholders.

By the end of this chapter, readers will have a clear understanding of the planning process for virtual power plant projects and be able to develop a comprehensive project plan that meets the project's objectives and requirements.

Chapter 7: Detail Design and Technical Implementation

Chapter 7 focuses on the detail design and technical implementation of the virtual power plant. It begins by finalizing the design specifications and detailed drawings based on the planning phase's output. This includes the hardware and software design for technical implementation, ensuring that all components and systems are properly integrated and optimized for performance. It also provides detailed instructions and guidelines for design and configuring the software components, including the control systems, communication protocols, and data processing algorithms [1].

It provides detailed instructions and guidelines for design and configuring the hardware components, including the energy generation devices, energy storage systems, and communication equipment.

Chapter 7 also emphasizes the importance of testing and validation in the technical implementation process. It describes the different types of tests that should be conducted, such as functional tests, performance tests, and reliability tests, and provides methods and tools for conducting these tests. By thoroughly testing the virtual power plant, any potential issues or problems can be identified and resolved before the plant is put into operation, ensuring its safe and reliable performance.

Chapter 8: Economic Analysis

This chapter introduce how to performs an in-depth economic analysis of the formal design to determine the composition and realization of the virtual power plant's construction and investment.

It begins by estimating the total project cost, including the capital expenditure (CAPEX) for equipment, installation, and infrastructure, as well as the operating expenditure (OPEX) for maintenance, operation, and management. The chapter also takes into account the financing options available for the project, such as loans, grants, and equity financing, and analyzes the impact of different financing strategies on the project's financial viability.

In addition to the cost analysis, Chap. 8 also examines the revenue streams of the virtual power plant, including energy sales, demand response payments, and ancillary services revenues. It develops a revenue model based on market projections and historical data, and evaluates the potential profitability of the project under different market scenarios. The chapter also presents a framework for assessing the return on investment (ROI) of the virtual power plant, taking into account factors such as the project's lifespan, discount rate, and inflation rate.

Furthermore, Chap. 8 addresses the economic risks and uncertainties associated with virtual power plant projects, such as energy price fluctuations, regulatory changes, and technological advancements. It provides risk management strategies and mitigation measures to minimize these risks and ensure the project's financial stability.

By the end of this chapter, readers will have a comprehensive understanding of the economic aspects of virtual power plant projects and be able to make informed decisions about the project's financial viability and investment potential.

Chapter 9: Future Development and Potential

Chapter 9 explores the future development and potential of virtual power plants.

It begins by analyzing the current trends and developments in the energy industry, such as the increasing penetration of renewable energy sources, the growth of the smart grid, and the emergence of new energy technologies. The chapter then examines how these trends are likely to impact the future of virtual power plants and identifies the opportunities and challenges that lie ahead.

Based on the trend analysis, Chap. 9 presents a vision for the future development of virtual power plants. It outlines the potential applications and use cases of virtual power plants in different sectors, such as residential, commercial, and industrial, and explores how virtual power plants can contribute to the transition to a low-carbon energy system.

The chapter also discusses the emerging technologies and innovations that are likely to shape the future of virtual power plants, such as artificial intelligence, blockchain, and the Internet of Things, and how these technologies can be integrated into virtual power plant systems to enhance their performance and functionality.

1.2 Definition of Virtual Power Plant

Virtual power plant [2] is an intelligent control technology and business model that aggregates the controllable resources on the load side that are originally invisible and uncontrollable in grid dispatching to form a controllable and tradable unit, directly participating in grid dispatching operations and electricity market transactions.

Virtual power plant is a cluster of dispersed energy converting installations, which are aggregated, that combines the controllable loads, power generation and energy storage of various distributed resources. It is a distributed power plant that can be operated independently.

Virtual power plants participate in grid dispatching operations and electricity market transactions with advanced technology. Their typical characteristics includes being digitalized entity, having diverse resources like controllable loads, power generation and energy storage without large hardware investment, and being a electricity market participant. Compared to traditional plants, they have flexible, diverse, eco-friendly resources and are competitive in power generation cost. They have various business mode and transaction roles, especially when cross-regional trading being a key feature.

It can aggregate various types of distributed resources such as controllable loads, distributed power generation units, and energy storage through the modern information and communication infrastructure such as intranet/internet on the basis of the traditional physical architecture of power grid. Then, it carries out intelligent coordination and optimized operation control targeting market transactions to achieve multi-energy complementarity on the power supply side and flexible interaction on the load side, in addition to innovative ancillary services such as power generation peak-shaving, power frequency-modulation, and reserved power generation capacity standby for the power grid dispatching.

1.3 Characteristics of the Virtual Power Plant

Compared with traditional power plants, virtual power plants have several different characteristics.

1.3.1 Virtual Versus Real Power Plant

A virtual power plant is not a real, physically existing power plant but rather a virtual one achieved through digital technology. However, from the perspective of the power grid, it functions the same as a real, physical power plant. A virtual power plant can generate electricity and also serve as an independent settlement unit, participating in power market transactions. During its operation, the virtual power plant follows grid dispatching instructions. By adjusting the increase or decrease in power generation, it participates in power regulation and load distribution within the power system. Additionally, it can contribute to peak and valley regulation in grid dispatching and system frequency modulation [3].

1.3.2 Extended Distributed Resources

The distributed resources included in the virtual power plant are expanded with many new features than the traditional power plant.

Energy storage devices, which can run in both directions, in the charging as a power load, and in discharging as a power generation device.

Controllable loads which can reduce power consumption according to demand at peak times and increase power consumption at low times. From the perspective of power dispatching grid, through the amount of absorbed power, controllable load can adjust the smoothness of the power generation and power consumption curves of the entire power grid, so as to participate in rapid demand response, thus making these innovative ancillary services for power market feasible.

The new additions to the virtual power plant have greatly enriched the means by which the virtual power plant participates in the operation of the power grid, and expanded the range of power generation capacity that can be regulated and the way in which it can be regulated.

1.3.3 Software Innovation Rather Than Hardware Investment

The planning and construction of virtual power plants does not require large-scale investment in new hardware facilities, but rather the use software innovation such as artificial intelligence, cloud computing and other up-to-date information technology, to tap the potential of flexible resources. In that, virtual Power Plant becomes a preferred technology option to quickly and efficiently improve the resilience of the power system to adopt irregular clean power energy and robustness of the power supply against power load fluctuations [4].

Therefore, decades after the emergence of the theory and application of virtual power plant, with the leapfrog progress of artificial intelligence technology, high performance computing, and the popularization and maturity of cloud computing, virtual power plant shows a huge potential in dealing with the new challenges to new energy and new power grid and gain more attention.

1.3.4 Benefits in Transaction

As an independent single market participant with trading authority, a virtual power plant aggregates small and dispersed small and medium-sized distributed power generation devices, energy storage devices, and controllable loads in a competitive power transaction platform. Under effective metering and settlement methods, based on a credible communication architecture, the virtual power plant participates in the market transaction on behalf of the aggregated distributed power resources mainly through real-time market bidding, thus obtaining bulk bargaining prices.

The virtual power plant is able to obtain preferential power sales prices through bulk bargaining, and at the same time participate in demand-side management and demand-side response, and obtain preferential power sales or direct value incentives based on capacity by improving the performance of the power grid [4].

1.3.5 Virtual Power Plant Versus Microgrid

Virtual power plant share many similar characteristics with microgrid, such as being able to participate in competitive market transactions as a single electricity market trader, but they are different. Virtual power plants are geographically dispersed, and the distributed generation resources that are subordinate to virtual power plants are usually small and have widely varying operating characteristics.

The virtual power plant is based on a communication system that aggregates dispersed available distributed resources, establishes a virtual network among the distributed generation resources belonging to it, and constitutes a virtual power plant, and does not participate in the investment and construction of equipment and materials such as the physical transmission and distribution network, and its operation and management. In contrast to virtual power plants, microgrids are geographically centralized and need to manage not only the distributed generation devices connected to the microgrid, but also the operation and maintenance of the microgrid.

1.4 Comparison of Virtual Power Plants with Traditional Power Plants

Compared with traditional power plants, virtual power plants have several different characteristics.

1.4.1 Diverse Resource Composition

VPPs can generate electricity by integrating various distributed generation sources. Additionally, they achieve energy savings and reserves by controllable demand loads through measures such as time-of-use pricing, interruptible tariffs, and user-side energy storage.

1.4.2 Environmentally Friendly Resources

VPPs primarily utilize new renewable energy sources, minimizing environmental impact. Power generation achieved through energy-saving technologies or demand load regulation even achieves zero emissions, with no adverse environmental effects.

1.4.3 Operational Synergy

Since electricity demand curves vary across regions, the cost of VPP power output differs geographically and temporally. To effectively participate in the electricity market—fulfilling peak-shaving, energy-saving, and emission-reduction roles—multiple VPPs must coordinate and collaborate across different operating entities.

1.4.4 Market Competitiveness

VPP-generated electricity can be safely dispatched and compete with conventional power plants in the electricity market.

1.5 Challenges to Virtual Power Plants

1.5.1 Low Carbon and Green Power

Driven by the background and goals of carbon neutrality and carbon peaking, vigorously promoting the green transformation of energy requires the construction of a new type of power system that is cleaner, lower carbon, safe and controllable, flexible and efficient, intelligent and friendly, and open and interactive with new energy sources, supported by source-network-load-storage interactions and multi-energy complementary integration, and centralized power generation and distributed power generation structures. Subject to light, wind speed and other environmental impact of photovoltaic power generation, wind power generation and other new energy generation devices can not continue to provide stable power to a large number of access to the power grid [5].

At the same time, with the improvement of the living standard of the residents, a large number of life-improvement power loads are connected to the system, including electric water heaters, air conditioners, kitchen appliances and other new types of power loads have increased dramatically, and a lot of changes and fluctuations will occur as a result of the influence of ambient temperatures and other meteorological adjustments.

The maturity and rapid promotion of new transportation technologies, such as new energy electric vehicles, has resulted in a large number of energy storage devices randomly connected to the grid. This has resulted in a pattern of new energy access to the grid, including wind power and photovoltaic power generation, a rapid increase in new energy vehicle charging piles accessing the grid, and a large number of random unplanned power loads accessing the grid, which has made the original power system even more complex, and the power system's security scheduling capability has been subjected to a huge challenge. In the case of global warming trend is obvious, abnormal climate characteristics are becoming more and more frequent, there has been a serious impact on industrial production, residential life of power scheduling problems.

1.5.2 Data Security and Privacy

Power operation data is highly sensitive, not only because of the importance of energy security itself, but also because of the large number of distributed resources oriented towards the public's social behavior, habits, and other matters of data discovery, access to a large amount of information related to security [6].

The technology of virtual power plant relies on advanced communication technology, and it is necessary to learn from the lessons learned in the development of the information industry, to prepare for the rainy days, and to deal with a large number of sensitive matters related to information security and privacy management in a proactive and compliant manner in the development of the virtual power plant. In order to ensure the efficient operation of the virtual power plant at the same time, the realization of legal compliance, including privacy protection, to prevent information leakage of information security.

1.5.3 Digital Technology

The virtual power plant breaks the physical constraints between power generators and consumers in the traditional electric power system, and makes full use of high-performance communication, reliable metering, and advanced control technologies to aggregate different types of distributed power generation resources.

Therefore, the virtual power plant does not need to transform the distribution network and other infrastructure, it can realize the distributed power generation energy to the utility grid stable transmission, and to provide rapid response services, and thus can significantly reduce the distributed power generation and the impact of the grid and the adverse effects.

1.5.4 Business Mode and Financial Incentives

A virtual power plant is an entity that realizes value creation and value realization by participating in demand-side response and demand-side management while taking part in power generation scheduling, including distributed power generation resources such as distributed power generation, energy storage devices, and controllable loads, as well as in bidding and trading in the power market.

Financial Incentives for VPP Stakeholders:

- For consumers, earn payments for reducing usage during peak hours, gain additional income from selling stored or surplus energy, and lower electricity bills through optimized self-consumption and time-of-use arbitrage.
- For Grid Operators and Utilities, reduce infrastructure costs by deferring grid upgrades through demand-side flexibility, enhance grid reliability, meet renewable energy targets by integrating more distributed renewables efficiently.
- For Investors and VPP Operators, multiple revenue streams from energy markets including ancillary services, and scalable business model with low marginal costs, renewable subsidies, and smart grid investments.

Virtual power plants represent a next-generation energy business model, combining digital intelligence, distributed energy assets, and market participation to optimize grid operations while unlocking new financial opportunities for all stakeholders. By leveraging flexibility, storage, and demand-side participation, VPPs enhance grid resilience, decarbonization, and economic efficiency in modern power systems.

1.6 Internationalization and Terminology

Virtual power plant as a new digital application technology of power technology cross-fertilization [2, 7], and a variety of power technologies are interrelated, the accurate use of standard terminology, which helps to correctly understand the concept of virtual power plant, and successfully participate in international exchanges. Through the relevant definitions, the concepts related to the operation of the virtual power plant can be determined, and the scene of the operation of the virtual power plant can be constructed.

The rapid development of virtual power plants is also reflected in the active collaboration including various international standards working groups. According to the consensus reached so far, the basic components of a virtual power plant include four parts: the generation side, the storage side, the user side, and the central control unit, which work in concert with each other to realize the basic functions of ensuring the supply of electric energy and ensuring the quality of electric power.

As a single important function for the existence and operation of the virtual power plant, it also includes the metering unit and the communication system. Through the metering unit, accurate metering can be achieved, optimized calibration, as an important settlement point, to participate in the power market transactions.

The communication system has various forms and flexible configurations, realizing reliable and trustworthy fast and timely processing of massive measurement sampling and control data, and aggregating a large number of distributed resources aggregated by the virtual power plant, truly as a unified whole, realizing the same efficacy as the physical power plant.

1.7 Summary

A Virtual Power Plant (VPP) aggregates distributed energy resources (DERs). It optimizes energy generation, storage, and consumption in real time, balancing grid demand and supply, to enhance grid stability, reduce costs, and support renewable integration without centralized infrastructure.

With the rapid development of clean energy, a large amount of clean energy can not be consumed, the traditional "source follows the load" development mode is unsustainable, and the construction of a new type of power system that interacts with the "source grid, load and storage" has become an inevitable trend.

The emergence of 5G technology, its wide connectivity, large bandwidth, low latency, high security features and power system massive resources agile response, accurate control requirements match, for the development of virtual power plants to provide technical support. Network deployment of 5G network empowers 700,000 square meters of Pinggao Industrial Park, realizing rapid construction without stopping work or production, meeting the network environment of 1000 modules and 10,000 devices accessing at the same time, and providing a highly reliable and secure information network for power transactions through 5G power soft and hard slicing technology.

Construction of energy facilities: Utilizing the idle resources of each subsidiary to build distributed clean energy and charging facilities, and configuring different energy storage facilities such as power storage, cold storage, heat storage, etc., to create a park-level "source-network-load-storage" energy network.

Platform construction: Comprehensively utilize 5G, fiber optic, NB and other communication methods to form a park-level energy information network, and connect each link of "source-network-load-storage" to the virtual power plant cloud platform in a hierarchical manner.

Improvement of energy data management: The energy data of the park has realized the three-level leap of second-level sensing and calculation, ten-thousand-level equipment access capacity, and multi-level eco-sharing, and achieved the optimal distribution of energy.

References

1. Yao, F. (2023). Engineering on the basis of big data in virtual power plant. In *Proceedings of 24th conference on electric power supply industry*. Xiamen, China.
2. *IEC TS 63189-1: 2023 Virtual power plants—Part 1: Architecture and functional requirements*. https://www.en-standard.eu/iec-ts-63189-1-2023-virtual-power-plants-part-1-architecture-and-functional-requirements/
3. Zhao, F. Y., Han, F., Qi, Y., et al. (2020). *Theory and practice of integrated smart energy*. China Electric Power Press.
4. Zeng, M., et al. (2016). Coordinated optimization operation mode and key technologies of "source-grid-load-storage" in energy internet. *Power System Technology, 40*(01), 114–124.
5. China Huadian Engineering Group Co., Ltd. (Ed.). (2025). *Clean energy and green computing synergy*. China Electric Power Press.
6. Lee, D., Park, D., Park, J. B., et al. (2016). Security-constrained unit commitment considering demand response resource as virtual power plant. *IFAC-PapersOnLine, 49*(27), 290–295.
7. *IEC TS 63189-2: 2023 Virtual power plants—Part 2: Use cases*. https://www.en-standard.eu/pd-iec-ts-63189-2-2023-virtual-power-plants-use-cases/

Chapter 2
Comprehensive Conceptual Introduction to Virtual Power Plants

Abstract Technically, virtual power plants consists of distributed power resources, controllable loads, and energy storage devices. It consists of a control center (with different functions under various architectures), metering units, and communication systems. It aggregates electric energy resources, solving the problem of managing scattered and irregular power sources. In terms of business model, diverse business models exist like peak-bottom electricity fee policy and demand side response. In terms of market transactions, virtual power plants face challenges due to the need for efficient and reliable power trading, and the need for flexible management of power resources. An electricity market with their access is essential, and they can conduct cross-regional trading with flexible management models. Overall, a virtual power plants is an aggregation of evolving technologies aiming to optimize power management and realize benefits.

2.1 Features of Virtual Power Plant

From the perspective of grid management and electricity trading market, a virtual power plant that constructed in accordance with similar norm and standards is no different from those of a physical power plants [1]. However, the virtual power plant is constructed to meet the requirements of the changing market conditions, such as the rapid development of renewable energy resources, the changing consumption habits of end users, and the need for flexibility in power distribution. The virtual power plant is constructed to be eventually put into operation, and to meet the regulations of the power trading market as well as the requirements of grid management [2].

At the same time, a virtual power plant is a special kind of power plant that does not have a specific geographic location, but has the ability to flexibly control a large number of distributed energy resources. The specific method is to unify the management of a large number of dispersed arrangements of small and medium-sized energy resources through advanced communication and large scale computing, by smart scheduling, intelligent control, and data analytics.

F. Yao, *Engineering Guide to Planning and Design of Virtual Power Plant*,
https://doi.org/10.1007/978-981-95-2231-6_2

In the long run, the virtual power plant can realize the management of distributed energy resources, and realize the optimization of power generation, load balancing, and energy storage, so as to adapt to the changing needs of end users.

The virtual power plant can realize the aggregation of distributed energy resources dispersed in the grid (i.e., distributed power generation, controllable loads and distributed energy storage), in order to become a unified scheduling of the "power generation system", and then participate in the demand-side response, and also participate in the operation of the regional system scheduling.

Unlike traditional physical power plants, which often have only one major form of energy generation, virtual power plants contain three types of facilities: distributed generation, energy storage, and controllable loads, and can at least realize new functions such as power generation forecasting, load forecasting, power generation and consumption time-sharing plan development, controllable load management, control and management of energy storage, coordination and optimization of distributed power generation, condition monitoring and communication, data acquisition, and data analysis.

Based on this, reliable communication management and intelligent coordination and control are the two technological cornerstones of virtual power plant construction, thus realizing the value re-creation by means of digital technology.

In summary, the operation of a virtual power plant is characterized by the following new features:

1. Combination of diversified resources.
2. Open system to the outside world, with the flexibility to add or remove new resources.
3. Interactive and coordinated operation.
4. Dual identity of power source and load at the same time.
5. Comprehensive generation prices are competitive in the electricity market.

The core function of the virtual power plant is still a power plant, is highly intelligent, highly flexible adaptive power plant, and has the ability to adapt to the changing needs of end users and the environment. Virtual power plants are not limited to a uniform geographical location and site, and can even cross different geographical areas, distributed in different locations in the power grid at multiple points.

Because the virtual power plant has a high degree of adaptability and flexibility, through the virtual power plant access to the grid, can innovatively solve the problems of grid development encountered, including new energy consumption, integrated energy services, grid ancillary services and many other new issues.

2.2 Resource Aggregation of the Virtual Power Plant

Virtual power plant[1] is a kind of advanced information and communication technology that realizes the aggregation, coordination and optimization of various distributed power and energy resources, including distributed power sources, controllable loads, and energy storage devices.

2.2.1 *Composition of the Virtual Power Plant*

The virtual power plant has a diversity of resources [2]. The virtual power plant can not only realize power production by combining a variety of distributed generation (DG) resources, but also realize energy-saving reserves by regulating controllable loads, adopting time-sharing tariffs, and energy storage during peak-load hours.

2.2.2 *Distributed Power Generation*

Distributed generation normally refers to clean energy distributed power generation devices that are near the power load and meet environmental requirements. Its installed capacity can be varied from a number of kilowatts, tens of megawatts, to a few gigawatts.

The power generation system mainly includes two types of distributed generation: domestic distributed generation (DDG) and public distributed generation (PDG). The main function of DDG is to satisfy the user's own load, and if there is a surplus of power, the excess power will be transmitted to the grid; If there is a surplus of power, the excess power will be delivered to the grid. If there is a shortage of power, the grid will provide power to the users. Typical DDG systems are mainly small-scale distributed power sources serving individual residential, commercial or industrial distribution, etc.

PDGs mainly deliver the power they produce to the grid and operate to sell the power they produce. Typically, PDG systems mainly contain new energy generation devices such as wind power and photovoltaic.

With reference to the current domestic clean energy power generation technology, specifically such distributed power include gas distributed, biomass, distributed (scattered) photovoltaic (wind power), geothermal energy, new hydrogen, tidal energy and other forms of energy distributed power generation resources.

[1] The definition adapts from IEC TS 63189-1:2023 Virtual power plants—Part 1: Architecture and functional requirements.

2.2.3 Controllable Loads

Controllable loads refer to loads that can be adjusted in real time according to energy demand under the control of a virtual power plant, such as charging electric vehicles and demand response loads with household controllers installed.

2.2.4 Energy Storage Devices

An energy storage device is a device that can convert electrical energy into other forms of energy, store energy when there is sufficient power, and release it in the form of electric energy when the grid needs it.

Energy storage devices can compensate for the fluctuation and uncontrollability of renewable energy power generation, adapt to changes in power demand, improve the weakness of the power grid caused by renewable energy fluctuations, enhance the system's ability to accept renewable energy power generation and improve energy efficiency.

Energy storage devices include distributed energy storage and centralized energy storage. New energy electric vehicles that can charge and discharge in both directions can be classified as energy storage devices.

2.2.5 Control Center

The control center refers to the combination of software systems and hardware resources at the core of the virtual power plant, and the control center realizes different functions under different control architectures.

Under the centralized control architecture, the control center directly controls the resources connected in the virtual power plant.

Under the architecture of distributed control, the control center can be further divided into a high-level control center and a low-level control center, and the low-level control center and the high-level control center are connected, participate in optimization decision-making, and be responsible for coordinating and controlling the local load and local distributed generation resources to which the underlying control center belongs.

In a fully distributed control architecture, the control center is simplified to a data exchange and data processing center, and each distributed unit performs the functions of optimal decision-making and coordinated control independently.

2.2.6 Metering Unit

Communication systems must be able to securely and smoothly ensure the exchange of information between the control center and hardware connection units. Virtual power plant using two-way communication system, not only can receive the current state information, but also can send control instructions to the controlled object signal. The communication system is an important link in the virtual power plant for energy management, data acquisition and monitoring, and communication with the power system dispatch center. Through information interaction with the power grid or with other virtual power plants, the management of the virtual power plant is more visualized, which facilitates the monitoring and management of the virtual power plant by the power grid.

Virtual power plant can be flexibly connected to distributed power generation, energy storage devices, controllable loads and other resources, do not need these parts of a close electrical connection between each other, so a unified communication system and control technology to achieve the grid scheduler or power trading managers and other participants to achieve the task of scheduling, management and other tasks.

2.2.7 Resource Aggregation in a Virtual Power Plant

As shown above, the virtual power plant realizes the aggregation of electric energy resources. Not only can include large-scale power generation, large power consumption units, but more importantly, for the dispersed, a large number of irregular power energy resources, providing a scientific management, orderly regulation options [3].

For the existing power system, the power grid has invested huge resources to establish a dispatch center for a large number of power plants for scheduling and security management. The emergence of new energy sources, the emergence of distributed energy sources, irregular changes, so that the power system very much dislike these new scattered slow small. Virtual power plant technology using new technical means, high-speed communication, effective computing, so that these resources are aggregated to form a large poly fast, not only to solve the existing challenges, but also the formation of a new pattern.

From this perspective, the virtual power plant is not a technology, but an aggregation of a class of technologies, and a rapidly evolving technology. Understanding and evaluating virtual power plants requires an analytical understanding of the technologies employed and the ability to evaluate the effectiveness in order not to be intimidated by virtual power plants. To give an exaggerated example, in other words, the earliest power systems connected to the grid required a handwheel, with the chief of duty looking at the meter and constantly selecting the handwheel in his hand to regulate the voltage and phase of the power plant to synchronize with the main grid. Power plants that use this technology exclusively can also be connected to a virtual power plant, if they are fast enough to make good choices instantly. Virtual power plants are extremely inclusive, and it is not the case that the mention of a virtual power plant is an advanced technology, it is not. It can also be a very poor and outdated technology, but solves real problems. Most importantly, resources are aggregated and benefits are realized. The most important and confusing thing about advanced technology is that it improves effectiveness and efficiency.

2.3 Business Model of Virtual Power Plants

The governing regulation and policy, electricity dealing market, and power-Grid management strategy altogether are prerequisite of development of It is necessary to give general scenery.

There are diversified business modes in virtual power plants, such as simple peak-bottom electricity fee policy, Demand Side Response business mode, generation capacity standby compensation. Although there topics are irrelevant to the title of this book, there are still many people could not understand how IT technology can bring profit without zero kWh electricity generated. Business mode shall make virtual power plant feasible, if not so, great loss incurred.

2.3.1 Electricity Market Transactions

In different countries, power trading markets are formed in parallel with the power grid. It is divided into two major segments, power sales and power purchases. The traditional technology is a single-ended grid with a one-way flow of power from generation to consumption. Such power market transactions have been running for many years and have a wealth of experience.

Virtual power plant power market transactions, a very important point is that the beginning of the emergence of double-ended grid, that is, the end of the use of electricity also has a power generation power supply, there is the ability to send power to the grid. For the traditional grid, this is unacceptable and is managed directly by prohibitions. Another problem is that the diversity of power resources allows for huge irregularities and unpredictability in both generation and use of power. For conventional grids, this one is again unacceptable, and the grid directly adopts the prohibition of power abandonment on the grid.

2.3.2 Electricity Market Trading with Virtual Power Plants

An electricity market with virtual power plant access is a necessary condition for the construction of a virtual power plant. Being able to provide user rights for power market transactions, as well as electricity price settlement is a basic condition.

In the access to the power market requirements, often need to accept the power market regulation, to meet the power quality requirements, to ensure that the quantity of power requirements, to comply with the requirements of power security.

The basic function of the power trading market of the virtual power plant is to restrain and manage the guaranteed quantity of electricity, guaranteed price of electricity, excess quantity of electricity and excess price of electricity. At the same time, in the region with the conditions, for power ancillary services can provide pricing and trading.

2.3.3 Cross-regional Power Market Trading of Virtual Power Plants

Virtual power plant has the biggest difference is not limited to the region, with the ability to cross-regional power trading. This is a basic requirement for virtual power plants to have.

This situation of virtual power plants needs to be solved in accordance with the classification of zoning. In the same management area, a single trading body can be used, using the same pricing system. In different physical partition regions, access to management units and reporting units. In the cross-region, internal segregation of virtual power plants is performed, and the corresponding settlement is performed according to different algorithms. Corresponding to the dynamic access users, who can freely choose to access the grid, can use the dynamic management method.

Such a model shows the flexibility of the virtual power plant even more. There are not many market areas with such conditions, usually in fully competitive market areas. The general advice is to be relatively stable and avoid fast settlements.

2.4 Typical Examples of Virtual Power Plants

Accordingly, the virtual power plant is a new generation of power system, and the development of the virtual power plant is a new generation of power system. There are many typical examples of virtual power plants in the world [3].

The *Next Kraftwerke*[2] is a virtual power plant company based in Germany. It is a subsidiary of the German power company *Deutsche Telekom*.

Next-Kraftwerke's virtual power plant business can be divided into three models: energy aggregation for the generation side, flexible energy storage supply for the grid side, and demand response for the demand side.

The Virtual Platform Provides Services to Renewable Energy Power Generation Enterprises

Due to the random and fluctuating nature of renewable energy generation, generators are often unable to deliver previously promised electricity to the TSO on time, and when this happens, the generator will incur all the costs required to balance the volume, such as buying expensive electricity from other generators. Virtual power plants can help power generators monitor power generation in real time, avoid inaccurate predictions of power generation, and save power generators costs. In addition, the virtual power plant monitors renewable energy prices in real time, assists power generators in optimizing the power product mix, helps power generators expand profits, and then earns ancillary income for aggregators.

The Virtual Platform Provides Short-Term Flexible Energy Storage Services to the Grid Side

After receiving a signal from the grid operator to increase or decrease the power generation, the central control system of the virtual power plant transmits this signal to the individual dispatchable renewable energy power plants, taking into account the constraints of response time, charging station capacity, power generation, etc., to adjust the power generation to support the grid frequency and offset the fluctuations caused by other units in the virtual power plant (photovoltaic solar and wind). Virtual power plants earn revenue by providing peak shaving, frequency modulation services from the generation side to the TSO.

The Virtual Platform Serves the Grid Side by Controlling the Electricity Consumption on the Demand Side

Since the increase of power generation and the decrease of electricity consumption are consistent with the peak regulation and frequency regulation effects of the power grid, the virtual platform can provide ancillary services to the power grid side through the control of the demand side, and then earn auxiliary fees. In addition, virtual power plants can reduce the cost of purchasing electricity by distributing consumption on the grid side to low-price periods in the spot market.

Next Kraftwerke's VPP demonstrates how digital aggregation and intelligent control can unlock the full potential of distributed energy resources. By leveraging "The Power of Many" it provides a scalable, market-driven solution for renewable energy integration, grid stability, and energy transition.

---

[2] Refer to the website https://www.next-kraftwerke.com/vpp/case-studies.

2.5 Summary

In summary, virtual power plants are emerging as a new generation of power systems, with the potential to transform the way we consume and generate energy. By leveraging digital aggregation and intelligent control, virtual power plants unlock the full potential of distributed energy resources and provide a scalable, market-driven solution for renewable energy integration, grid stability, and energy transition.

References

1. Zhao, F. Y., Han, F., Qi, Y., et al. (2020). *Theory and practice of integrated smart energy*. China Electric Power Press.
2. *IEC TS 63189-1: 2023 Virtual power plants—Part 1: Architecture and functional requirements*. https://www.en-standard.eu/iec-ts-63189-1-2023-virtual-power-plants-part-1-architecture-and-functional-requirements/
3. *IEC TS 63189-2: 2023 Virtual power plants—Part 2: Use cases*. https://www.en-standard.eu/pd-iec-ts-63189-2-2023-virtual-power-plants-use-cases/

Chapter 3
Control Architecture of Virtual Power Plants

Abstract The virtual power plant's control architecture is vital for power market participation and grid scheduling, mainly categorized into three modes. In the centralized control mode, the central control center has full control over all units, but it faces high communication and computational loads with limited scalability due to repetitive information. The distributed control mode uses hierarchical control, where local control centers manage regional units, formulate plans, and exchange information with the high-level center, improving communication efficiency and scalability. The fully distributed control mode consists of autonomous subsystems with intelligent agents. The central center acts as a data hub, providing excellent scalability and flexibility, though with slightly weaker real-time terminal resource management, and enabling large-scale, diverse management.

Virtual power plant as an independent entity to participate in the power market transactions and grid scheduling, therefore, the virtual power plant within the various resources, need to be used to with the appropriate Control Architecture, flexible collaboration, scheduling and control. According to the different control structure of virtual power plant information flow transmission, the control mode of virtual power plant can be divided into: centralized control mode, decentralized control mode, and completely decentralized control mode.

3.1 Centralized Control Structure

Centralized control [1] implements complete control over all power generation units and power consumption units within the virtual power plant, and assigns scheduling plans through the centralized control center. The objects of centralized control include distributed generation units, controllable loads, and energy storage devices, to be controlled with central control units, power metering units, and communication systems [1].

With the centralized control architecture, all information within the virtual power plant needs to be transmitted to the central control center, which has control over all units and specifies the generation or consumption plan for each unit. The central

F. Yao, *Engineering Guide to Planning and Design of Virtual Power Plant*,
https://doi.org/10.1007/978-981-95-2231-6_3

control center has a high level of control authority, flexible control measures, and the best optimization based on big data.

Under such an architecture, the communication system central operates at a high full load rate and control center has a large amount of computation. At the same time, the measurement and control information and control instructions of the said power generation units and load units need to be uploaded and downloaded frequently, with a lot of repetitive and invalid information. The scalability of the virtual power plant is easily constrained by the computation capacity of the central control center after the completion of the construction of the virtual power plant.

The centralized control architecture is shown in Fig. 3.1.

3.2 Distributed Control Architecture

Distributed control [1] implements hierarchical control. Taking the two levels of control including the high-level control center and the local level control center as an example, the high-level control center of the virtual power plant assigns control tasks to the local level control center, and the local level control center is responsible for formulating power generation plans and power consumption plans for the

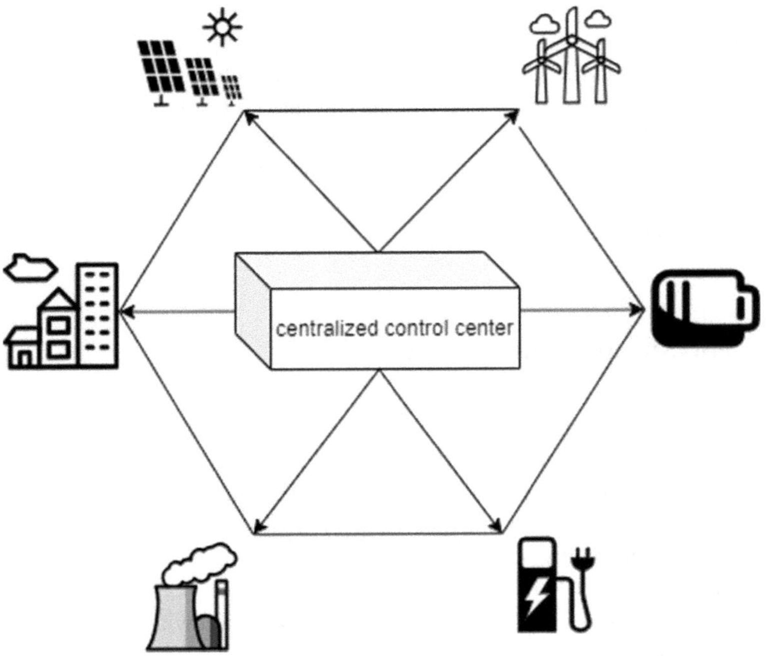

Fig. 3.1 Centralized control architecture

power generating units and power consuming units under its control according to the instructions of the control tasks.

The control architecture of distributed control includes a power metering unit, a communication system unit, a high-level control unit, an on-site control unit, and all distributed power generation units, controllable loads, and energy storage devices connected to the on-site control unit. In the distributed control architecture, the lowest level local control unit not only needs to have the communication capability to receive the control instructions from the high-level control center, but also needs to have the control function to perform the control function for the lowest level power generation unit and power consumption unit.

In the distributed control architecture, the local control center manages only a limited number of power generation units and power consumption units in the region, is responsible for formulating the power generation plan and power consumption plan of each unit, exchanges information within a limited range, and transmits the aggregated information to the high-level control center. The high-level control center decomposes and distributes the tasks to each local control center. Such a structure helps to improve the problems of data congestion and communication efficiency degradation in the centralized control architecture, and improves the scalability of the whole system.

The distributed control architecture is shown in Fig. 3.2.

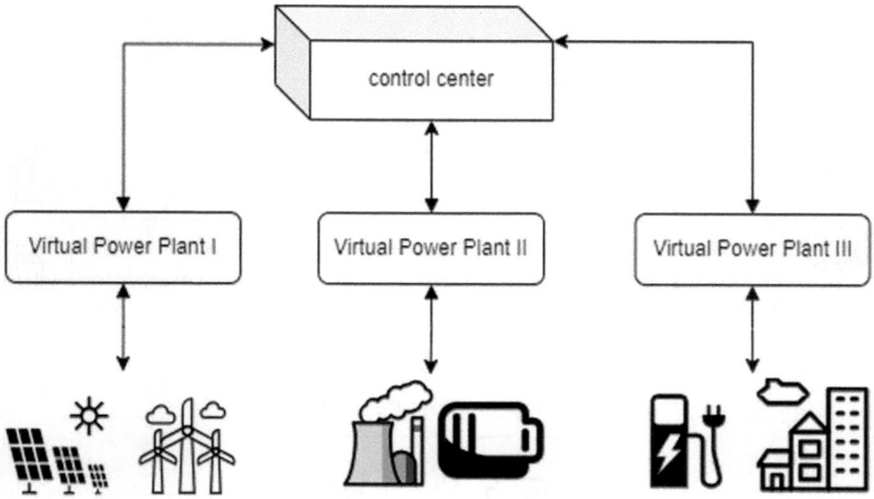

Fig. 3.2 Distributed control architecture

3.3 Fully Distributed Control Architecture

The fully distributed control architecture [1] consists of multiple mutually autonomous, independent intelligent virtual power plant subsystems, each of which realizes the tasks originally realized by the central control center through its own intelligent agent. The central control center of the virtual power plant only serves as a data exchange and data processing center, connecting each virtual power plant subsystem to provide data exchange and data processing services. The fully distributed control architecture includes power metering units, communication systems, a central data exchange and data processing center, intelligent agents of the virtual power plant subsystems and the distributed power supplies, controllable loads, energy storage devices, etc., to which the intelligent agents are connected.

The fully distributed control architecture provides excellent scalability and configuration flexibility, so that the addition of new virtual power plant subsystems or new distributed generation resources does not affect the operation of the existing virtual power plant. The central data exchange and data processing center in the center can manage more data for data mining and intelligent optimization. For the terminal distributed generation resources, the scheduling and management of real-time is slightly poor. Fully distributed control architecture can realize a larger scope, even across a larger geographical area, and realize multiple forms of management mode.

The fully distributed control architecture is shown in Fig. 3.3.

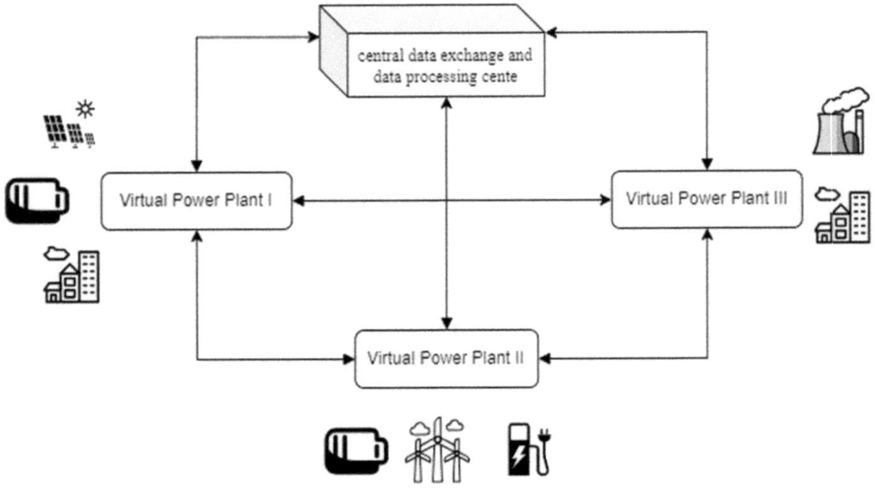

Fig. 3.3 Fully distributed control architecture

Reference

1. Yao, F. (2023). Engineering on the basis of big data in virtual power plant. In *Proceedings of 24th conference on electric power supply industry*. Xiamen, China.

Part II
To Professional

Dive deep for understanding on *Virtual Power Plant*, with details for professionals.

Chapter 4
General Part of the Engineering Guide

Abstract General Engineering Guide chapter on virtual power plants introduces the project workflow and features, which are based on the principle of virtual power plants. The scope of this chapter applies to VPPs with specific capacity ranges for aggregated resources. The workflow includes planning and design stages. Basic provisions emphasize safe operation, openness, and regional adaptability. Information security is crucial, with principles like protecting high-security systems and having reliable equipment and software. Overall construction principles are advanced, secure, open, data-integrity ensured, scalable, and standardized. Digital and AI technologies like cloud computing, AI algorithms, digital twin, etc., are mentioned for application in VPPs.

This chapter introduce the workflow of virtual power plant project and prominent features of virtual power plant. The workflow gives hand-on guide on engineering, professionals are capable of following the workflow to achieve their business goals. Artificial Intelligent and advanced communication technology shall be adopted in virtual power plant, by which revenue is generated. Planning stage and engineering design stage are two main block in this book.

4.1 Scope

In accordance with the guiding principles that define the planning and design techniques for virtual power plants, this applies to the planning and design of virtual power plants with a planning capacity of greater than 10–1000 MW for the overall aggregated resources and greater than 10–100 kW for the individual distributed aggregated resources. Other special requirements and application scenarios can be referred to.

F. Yao, *Engineering Guide to Planning and Design of Virtual Power Plant*,
https://doi.org/10.1007/978-981-95-2231-6_4

4.2 Terms and Definitions

Where the terms already exist in the normative citation documents, they are not repeated in this document[1].

Virtual Power Plant (VPP)

Groups of multiple Distributed Electricity Generation Resources (DERs) that are aggregated to act as a single controllable generation unit.

Distributed Energy Resource (DER)

A generating unit with a secondary protection system that is connected to a low or medium voltage power network and contains loads that can be placed in generation mode (e.g., electrical energy storage systems).

Distributed Generation (DG)

A type of power generation that utilizes multiple energy resources to generate electricity when connected to a low-voltage or medium-voltage power network.

Energy Storage Devices

Energy storage devices are devices that utilize mechanical, electromagnetic, electrochemical, and phase change energies to store, convert, and discharge recyclable electricity.

Controllable Loads

Controllable loads that can be controlled by the VPP to adjust the amount of generated electricity.

Demand Side Management (DSM)

Management of electricity consumption realized at the customer side. The management of electricity by an authorized administrator who directs consumers to use less electricity during peak times and more electricity during low times. It is a type of optimization that is achieved through government policy guidance or commercial agreements to improve electrical energy efficiency and optimize electrical energy consumption.

Demand Side Response (DSR)

The management behavior of changing the consumption behavior of electricity from the demand side according to changes in electricity prices or preferential measures, so as to smooth or shift the load curve, reduce the difference between the generation side and the consumption side, and improve the operational efficiency of the entire power system.

Electricity Market

Electricity market, in the usual sense, refers to the aggregate relationship between generation, transmission and distribution, consumption and sale of electricity. In particular, it refers to a competitive electricity market in which electricity producers and electricity users, through inquiry, bidding, or other methods, trade in electricity and related energy products. In this market, the confirmation of transaction prices and transaction quantities is realized through market competition.

4.3 Workflow and Roadmap

The process includes planning and design stages. It is executed step by step in accordance with the following workflow [2].

The planning stage begins with requirement analysis, where objectives such as grid stability, renewable integration, and demand response are identified, alongside defining the project scope—including geographical coverage, asset types, and scale—and assessing regulatory and market compliance.

This is followed by resource assessment, which involves inventorying distributed energy resources (DERs) like solar PV, wind, batteries, EVs, and demand-side loads, then evaluating their availability, capacity, and controllability. A feasibility study concludes the planning phase, examining technical feasibility (grid connectivity and communication infrastructure), economic viability, and risk assessment (cybersecurity and operational risks).

Moving to the design stage, architecture design outlines the control hierarchy—weighing centralized versus decentralized logic—and establishes the communication infrastructure, including protocols and network topology, along with selecting software platforms featuring aggregation algorithms, forecasting tools, and SCADA integration.

The DER integration strategy focuses on standardizing interfaces for diverse assets and defining participation rules such as priority dispatch and bid curves. Market and grid integration design includes participation models for energy markets and adherence to grid codes covering voltage/frequency regulation and fault ride-through. Cybersecurity and data management measures, such as encryption, access controls, intrusion detection, and a data architecture for real-time monitoring and historical analytics, are also integrated here.

The implementation workflow starts with pilot testing to validate control strategies using a subset of DERs, followed by scaled deployment across the full asset portfolio. Grid interconnection involves securing approvals and conducting interoperability tests, with commissioning ensuring performance aligns with design specifications.

Finally, operation and optimization involve continuous monitoring via IoT/cloud platforms, dynamic reconfiguration in response to weather, demand, or price signals, and leveraging machine learning to enhance forecasting and asset coordination.

4.4 Basic Provisions

4.4.1 Brief

Based on the traditional electric power business process, the virtual power plant integrating digital technology, the operation mode is complex, the management subject is diversified, and it is necessary to ensure safe and reliable operation, economic and technical reasonableness, in order to ensure the standardized management, improve the level of institutionalization, and solve the problem of standardization of the project.

4.4.2 Platform

The virtual power plant platform has the characteristics of openness and sharing, and the planning and design of the virtual power plant should meet the characteristics of open put and scalability. Distributed energy resources with access conditions, the virtual power plant platform should provide access methods and processes, and should also provide methods and processes to exit the virtual power plant platform.

4.4.3 Virtualization

The virtual power plant project has cross-regional, virtualization characteristics, virtual power plant planning and design should be clearly defined regulations to adapt to the management of the region, to comply with the requirements of the region's power scheduling and power management.

4.4.4 Guiding Principles and Regulations

This paper sets out the requirements for the basic guiding principles for the planning and design of virtual power plants, when in conflict with the provisions of national laws and administrative regulations, should be implemented in accordance with the provisions of national laws and administrative regulations.

4.4.5 *Information Security*

The virtual power plant is based on the application of digital technology in the field of electric power and energy, and needs to comply with the regulations and provisions on information security and data security. At the same time, virtual power plants cover a wide range of areas, the aggregated distributed resources are distributed over a wide geographical area, and their aggregated power energy data are inherently sensitive. Virtual power plants face information security challenges in an environment of increasing cybersecurity and information security risks on an international scale [3].

Compliance with relevant legal and regulatory requirements, selection of applicable technologies, and a more contextualized combination of technical measures ensure efficient data transmission and information security in virtual power plants. The information security of virtual power plant needs to protect the information resources such as hardware, software and data in the system from accidental or malicious damage, tampering and leakage, to ensure the normal operation of the system and uninterrupted service.

4.4.5.1 Basic Principles of Security Protection

The basic principles of virtual power plant information security protection are:

1. In the power system, systems with higher security levels are not affected by systems with lower security levels. The security level of the comprehensive monitoring and dispatching system is higher than that of the comprehensive management information system and office automation system, and each comprehensive monitoring and dispatching system must have its own security protection facilities with high reliability, and must not be directly connected with systems of low security level;
2. The integrated monitoring and dispatching system can be interconnected with other local monitoring systems of the same level through a dedicated LAN, or interconnected with upper and lower level off-site monitoring systems through the power dispatching data network;
3. The integrated system monitoring and control system shall not be connected to the Internet and the use of e-mail shall be strictly limited.

The information security of the information system of the virtual power plant shall support system alarms and a complete security event alarm mechanism, so that alarm information can be output to the third-party log alarm management system when illegal invasion, device abnormality, communication interruption or loss of application data occurs. When required by the local electric power management organization, its information security must support the device management system of the subordinate dispatching management organization to remotely manage the equipment and information within the scope of its applicable management.

4.4.5.2 Requirements to Be Met by Equipment and Software of the Information System

The equipment used in the information security system shall meet the following reliability requirements [4]:

1. It shall pass the electromagnetic compatibility test;
2. It shall support the function of dual-machine hot standby and automatic switching in case of failure of either equipment;
3. It shall pass the security review and technical appraisal organized by the Office of the State Cryptography Administration Committee and have the sales license of the Ministry of Public Security;
4. Should use equipment with intellectual property rights, the intellectual property rights of the equipment used in the jurisdiction of foreign entities, the need to provide a legally binding commitment to reliability.

Physical protection of information security requires the protection of information systems, hardware and software equipment and facilities, as well as other media from natural disasters, man-made damage and operational errors or influence. The site where the information system is deployed needs to have a convenient environment to prevent unauthorized personnel from entering and to facilitate security patrols and inspections. With anti-vibration and anti-seismic, lightning, fire, water and moisture, anti-static measures, in line with the requirements of electromagnetic protection, appropriate environmental temperature and humidity.

Information security software needs to have sufficient performance to ensure that at any time, are guaranteed to run reliably and stably, the computer code needs to be verified by the software to prevent code leakage, code invasion, to prevent automatic software upgrades.

The platform on which the information system software runs needs to be safe and reliable, and obtain valid authorization from the copyright holder. The operating system platform cannot be open to remote access and automatic upgrade functions. Backup security measures, including port scanning, need to be taken for system software to prevent the risk of remote activation of built-in modules.

Third-party modules and open plug-ins used by the information system need to be listed and logged, and the plug-ins need to be verified and periodically checked for new risks.

The part of the information system that uses the network needs to have network security measures to realize authentication, access control, firewall, intrusion detection and other network security measures.

The stringent requirements for information security in virtual power plants make it necessary for solutions using cloud computing technology to be specifically evaluated. When the public cloud solution cannot meet the requirements after evaluation, other technical solutions need to be adopted to solve the problem of information security.

4.5 Overall Construction Principles

4.5.1 *Advanced*

The virtual power plant is based on the application of digital technology in the field of electric power and energy, and needs to comply with the regulations and provisions on information security and data security. At the same time, virtual power plants cover a wide range of areas, the aggregated distributed resources are distributed over a wide geographical area, and their aggregated power energy data are inherently sensitive. Virtual power plants face information security challenges in an environment of increasing cybersecurity and information security risks on an international scale.

4.5.2 *Security*

The database system used in the system database, to ensure that the original data of electric energy can not be modified, the model of measuring and settlement of electric energy in the corresponding derived library, derived library data can only be modified under the authorization of permission, the establishment of perfect security measures for different levels of users, the establishment of the appropriate access rights to ensure that the legitimacy and seriousness of the electric energy and billing. At the same time, the system supports automatic or manual data archiving and recovery.

4.5.3 *Openness*

The system has sufficient open performance, the software system has been reserved in the interface and function, only through a simple configuration, can allow different manufacturers of products to form a complete system, and through a wealth of built-in software interfaces (OPC, DDE, ODBC, etc.) and third-party systems seamlessly integrated to provide low-cost IBMS integrated management solutions.

4.5.4 *Data Integrity*

Due to the cumulative and transmissible characteristics of electric energy data, it is required that the original data of electric energy should not be lost under any circumstance, which is especially important when carrying out segmented and subrate electric energy statistics and settlement. In this system, a variety of technical means are used in the collection, processing and transmission to ensure data integrity.

4.5.5 *Scalability*

The bus capacity, software resources, module IO point configuration in the system program have a certain margin, so as to flexibly increase a small number of control points according to the owner's requirements without additional costs. The system design adopts networked structure, giving full consideration to the user's future expansion of the energy center and the need for functional expansion, which can be easily achieved by increasing the local collection of instrumentation, but also through the network expansion, expansion of the new control network bus, the system scale can be multiplied.

4.5.6 *Standardization*

The key hardware device of this system is the data gateway, which is safe, reliable and corresponds to all mainstream metering instruments. The main features are: the data gateway should support cycle mode data collection, fixed moment data collection and current moment data collection, and can accept the commands and related settings issued by the data center through the data management platform.

4.6 Application of Digital Technology and Artificial Intelligence Technology

The application of digital technology and artificial intelligence technology needs to be arranged gradually in the process and correspondingly confirmed in detail [5].

Cloud Computing

Micro-service technology based on cloud-native architecture is applicable to a variety of scenarios for deployment.

Artificial Intelligence

Using algorithmic models to realize accurate prediction, scheduling plans and data analysis.

Digital Twin

Establish a virtual mapping of the real scene on the basis of a 3D model.

Blockchain

Decentralized blockchain technology for transaction applications.

Big Data

Storage, cleaning and utilization of the massive data collected by IOT, and the data is processed and analyzed by artificial intelligence.

Edge computing

Distributed, low-power intelligent gateway technology, based on the cloud-edge architecture, to realize the intelligent analysis of the data generated by the distributed edge computing node.

IoT

High-performance, multi-protocol, multi-scene, multi-device IoT acquisition technology.

References

1. *IEC TS 63189-1: 2023 Virtual power plants—Part 1: Architecture and functional requirements*. https://www.en-standard.eu/iec-ts-63189-1-2023-virtual-power-plants-part-1-architecture-and-functional-requirements/
2. *IEC TS 63189-2: 2023 Virtual power plants—Part 2: Use cases*. https://www.en-standard.eu/pd-iec-ts-63189-2-2023-virtual-power-plants-use-cases/
3. Lee, D., Park, D., Park, J. B., et al. (2016). Security-constrained unit commitment considering demand response resource as virtual power plant. *IFAC-PapersOnLine, 49*(27), 290–295.
4. Yao, F., Xie, G., Li, C., Liu, X., Ren, B., Lu, H., Li, R., Song, W., Duan, C., & Wang, W. (2024). Flexible construction technology and application of new integrated energy systems. In *Proceedings of the 2024 new power system development forum* (pp. 94–99). China Electric Power Press.
5. China Huadian Engineering Group Co., Ltd. (Ed.). (2025). *Clean energy and green computing synergy*. China Electric Power Press.

References

Chapter 5
Prerequisite Condition and Feasibility Determinationy

Abstract This chapter details the prerequisite and feasibility study for virtual power plant (VPP) projects. Success hinges on adhering to regulations, having a supportive market mechanism, effective operation strategies, load-generation balance, and proper infrastructure. Construction conditions are crucial and categorized into external, demand-side, and technical aspects. External conditions cover policies like national energy strategies and market operation mechanisms, with various regulations and market rules providing a framework. Demand-side conditions focus on grid operation, user safety, and resource consumption challenges, such as new energy integration and load fluctuations. Technical conditions require equipment to be intelligent, secure, and user-friendly, enabling safe access, automation, data integrity, and good user interaction. Commercial conditions related to load categories are also noted.

5.1 Brief

There are prerequisite for successful project construct, regulation, codes, business market mechanism, and electricity fee incentive policy for virtual power plant and power grid management. These determine the feasibility of virtual power plant project construct, including,

Regulation and codes for virtual power plant and power grid management.
Business market mechanism and electricity fee incentive policy.
Peak-load shift, power frequency adjustment, grid stabilization and operation safety strategy.
Load-generation balance for solar power, wind power, and other renewable energy sources, hydrogen generation, battery storage, and dual-mode electricity vehicles.
Infrastructure condition such as communication, automation, and data security.
Technical level and technical capability of equipment and devices.

F. Yao, *Engineering Guide to Planning and Design of Virtual Power Plant*,
https://doi.org/10.1007/978-981-95-2231-6_5

5.2 Construction Conditions and General Provisions

5.2.1 Scope

The prerequisite condition of virtual power plant projects should be investigated before planning and designing, and the construction conditions of virtual power plant projects should be determined to meet the needs of project construction. The construction conditions of virtual power plant projects include external conditions, demand-side conditions, and technical conditions [1].

5.2.2 External Condition

Construction conditions include external conditions, demand-side conditions, and technical conditions.

Through the analysis of construction conditions, the boundary conditions and constraints of the virtual power plant project are determined in order to provide competitive prices for power generation and power ancillary services, to realize safe operation, and to improve power quality.

The analysis of construction conditions, the important decision-making basis for virtual power plant planning.

5.2.3 Phasing

The project construction of the virtual power plant needs to meet the requirements of phased construction, dynamic management, and flexible adaptation to a variety of operating conditions.

5.2.4 Normative References

The documents outlined below form integral provisions of this regulation, incorporated through normative references that guide virtual power plant (VPP) design and operation. These references encompass a categorized set of standards, guidelines, and frameworks specific to VPPs, addressing technical, regulatory, and operational requirements. Importantly, the application of these documents varies by region, with local jurisdictions mandating the adoption of those that align with their unique grid codes, market structures, and compliance mandates.

Information System and Data Center Design

Specification for the Design of Computer Rooms for Electronic Information Systems.
Design Code for Comprehensive Cabling System Engineering.
Specification for Entrance and Exit Control System Engineering Design.
Interim Provisions on Design Management of Building Intellectualization System Project.

Power Plant and Grid Systems

Coding Standard for Power Plant Identification System.
Design Code for Large and Medium-sized Thermal Power Plants.
Technical Provisions on Information System Design for Thermal Power Plants.
Technical Regulations for the Design of Regional Power Grid ESC Automation.
Application Layer Protocol for Real-time Data Communication of Power System.
Technical Regulations on the Design of Power System Dispatch Automation.

Information Security and Cybersecurity

Guidelines for the Classification of Computer Information System Security Protection Levels.
Technical Requirements for Network Basic Security.
Technical Requirements for General Security of Information Systems.
Technical Requirements for Operating System Security.
Technical Requirements for Security of Database Management Systems.
Technical Requirements for Physical Security of Information Systems.
Basic Requirements for Information System Security Level Protection.
Information System Security Level Protection Classification Guidelines.
Basic Requirements for Network Security Level Protection.
Data Security Capability Maturity Model.
Guide to Big Data Security Management.
Technical Requirements for Data Security of Government Information Sharing.
Provisions on Security Protection of Electric Power Monitoring Systems.
Specification for Security Protection Programs and Assessments of Electric Power Monitoring Systems.

Software Engineering and Documentation

Guidelines for the Preparation of Computer Software Requirements Statements.
Software Survival Cycle Process.
Specification for the Specification of Computer Software Requirements.
Specification for the Preparation of Computer Software Testing Documents.
Software Engineering Terminology.
Guide to Software Document Management.
Guide to Software Maintenance.
Specification for Computer Quality Assurance Programs.

Computer Quality Assurance

Guidelines for the Preparation of Computer Software Testing Documents.
Software Testing Process.
Specification for the Specification of Computer Software Testing Requirements.

Network and Communication Standards

Network Technical Requirements Network Performance Parameters and Indicators (IP).
Information Format and Protocol for Serial Data Interface of Alarm Transmission System.
Management Requirements for Security Level Protection of Computer Information Systems.

Electrical and Equipment Safety

Safety of Information Technology Equipment Including Electrical Equipment.
Safety of Power Equipment.
Safety of Communication Equipment.

Emerging Technologies

Specification for the Management of Virtual Power Plants.
Technical Specification for Resource Allocation and Evaluation of Virtual Power Plants.

Computer and Information Technology

Guidelines for the Preparation of Computer Software Requirements Statements.
Software Survival Cycle Process.
Specification for the Specification of Computer Software Requirements.
Specification for the Preparation of Computer Software Testing Documents.
Software Engineering Terminology.
Guide to Software Document Management.
Guide to Software Maintenance.
Specification for Computer Quality Assurance Programs.

Computer Quality Assurance

Guidelines for the Preparation of Computer Software Testing Documents.
Software Testing Process.
Specification for the Specification of Computer Software Testing Requirements.

5.3 External Conditions

5.3.1 Scope

The external conditions of the virtual power plant constitute the external constraints for virtual power plant planning.

The external conditions mainly include the conditions of policies and regulations and the conditions of market operation mechanism.

5.3.2 Conditions of Policies and Regulation

The conditions of policies and regulations include the following.

1. Requirements of the national energy development strategy. This includes the dual-carbon goal of "peak carbon and carbon emission reduction", and the operational requirements of the new power system in the future in terms of "green, flexible, multi-interactive and highly market-oriented". Specific requirements include the provision of power demand-side response capability and the consumption of a variety of new types of power loads;
2. The implementation of the national energy development strategy is specific to the specific requirements of the industry and region, including but not limited to the power system reform and business transformation requirements of power grid enterprises, and the requirements for the construction of regional power grids to be intelligent and smart grids. According to the specific requirements, to realize the "source and load interaction", "source, network, load, storage synergy" and other needs;
3. Effectively solving the problems of energy saving and emission reduction, new energy consumption, grid peak and frequency adjustment, etc., and generating actual economic benefits and comprehensive synergistic benefits;
4. Requirements of demand-side management of electricity in different regions. According to the management system of user demand side management, the load resources on the demand side realize the optimized management of integrated resources and obtain special subsidies for demand side response.

5.3.3 Conditions of Market Operation Mechanism

The conditions of the market operation mechanism include the following.

Market Operation

The conditions of the market operation mechanism include the following.

1. The market environment that is compatible with the construction of the virtual power plant;
2. It shall be in line with the power market mechanism and market power trading rules of the region where the project is located;
3. It shall be consistent with the different stages of development of the electricity market in the region where the project is located. Functional requirements for power market transactions at various stages of development, such as invitation stage, marketization stage and autonomous dispatch stage, shall be provided;
4. It shall have a real-time market trading mode based on bidding strategy, and be able to adjust dynamically and update online with the development of market-based mechanism and the adjustment of market-based rules;
5. It shall have the requirement of realizing users' participation in demand-side response and responding to auxiliary power service capability;
6. It should meet the performance requirements of practicability, availability and effectiveness required for the operation of the electricity market;
7. The construction of the virtual power plant shall comply with the rules of marketization, have the ability to effectively control the construction costs, and have the ability to sustain profitability.

5.3.4 Regulations and Codes

The typical relevant regulations and conditions for virtual power plants shall be as follows.

Basic Rules for Electricity Market Operation

The Basic Rules for Electricity Market Operation define virtual power plants (VPPs) as legitimate market participants, allowing them to fully engage in the electricity market. The key provisions are as follows:

Electricity market participants include business entities, electricity market operators, and grid companies providing transmission and distribution services. Among them, business entities consist of power generation enterprises, electricity retailers, electricity consumers, and new types of market participants (including energy storage companies, virtual power plants, load aggregators, etc.). Electricity market operators include electricity trading institutions and electricity dispatch institutions.

The electricity market operates under a registration system. Electricity trading institutions shall establish a market registration mechanism in accordance with national regulations and be responsible for managing market registrations. Business entities entering or exiting the electricity market must complete the corresponding registration procedures.

Specification for the Management of Virtual Power Plants

The Virtual Power Plant Management Specification outline general principles for the entire lifecycle of virtual power plants (VPPs), specifying framework requirements for their planning and design, construction and implementation, grid connection testing, registration and integration, operation and management, decommissioning and deregistration, and archival documentation.

The document comprehensively considers factors such as timeliness, safety, and economic efficiency in VPP resource allocation. It provides allocation standards for various application scenarios, including peak shaving, demand response, frequency regulation, reserve capacity, and energy trading in power markets. Additionally, the document evaluates VPP resource allocation plans from technical effectiveness and economic benefit perspectives, enabling an objective assessment of the advantages and disadvantages of VPP resource composition. This serves as a guiding reference for determining the scale and application scope of VPPs.

Furthermore, the specificat delegate specific implementation requirements and detailed rules to local authorities, enhancing their feasibility and practical applicability.

Technical Specifications for Virtual Power Plant Resource Allocation and Evaluation

The Technical Specifications for Virtual Power Plant Resource Allocation and Evaluation define the resource allocation requirements for virtual power plants (VPPs). It provides guidance for optimizing VPP resource configurations across four key application scenarios:

1. Participating in power system peak shaving or demand response,
2. Engaging in power system frequency regulation,
3. Providing reserve capacity for the power system,
4. Participating in electricity energy markets.

On the other hand, the document establishes evaluation methodologies for VPP resource allocation, offering objective criteria to assess the effectiveness of different resource compositions. These guidelines play a crucial role in determining the appropriate scale and application scope of VPPs, ensuring their efficient integration and operation within the power system.

Detailed Rules for Virtual Power Plant Ancillary Services Market Transactions

The Detailed Rules for Electricity Ancillary Services Market Transactions formally recognize qualified virtual power plants (VPPs) as legitimate market participants in the ancillary services market, granting them a clear and official status. VPPs are permitted to engage in market transactions either directly or through authorized load aggregators by signing agency agreements.

During the initial phase of market participation, VPPs and controllable load resources will take part in the peak shaving market using a "quantity declaration without price bidding" mechanism and will be prioritized over thermal power in the clearing process. This simplified participation approach is designed to lower entry barriers and enhance VPPs' involvement in the electricity market.

The rules aim to streamline market access for flexible resources while ensuring system reliability and promoting the efficient integration of distributed energy assets into grid operations.

5.4 Demand-Side Conditions

5.4.1 Scope

The demand-side conditions for distributed resources [2] included in the scope of construction of the virtual power plant include the requirements for the operation and management of the grid to which the distributed resources are connected, the requirements for the safe and stable use of electricity by power users, and the requirements for comprehensive economic returns.

5.4.2 Power Grid Requirements

The demand-side conditions of distributed resources accessing the power grid specifically include:

1. Scheduling requirements for peak and frequency adjustment of the regional power grid to be accessed
2. The secure and stable power generation of the accessed regional power grid to meet the requirements of power quality standards and power availability;
3. the local load of the accessed regional grid, the requirements of spatial distribution imbalance;
4. The local load of the accessed regional power grid, the requirements of uneven distribution in time;
5. The requirements of the accessed regional power grid that the construction conditions and construction progress of supporting power equipment do not match the load growth trend.

5.4.3 Load Requirements

Unbalanced consumption of distributed resource types in the accessed regional grid. Specifically include:

1. The need to improve the utilization rate of new energy power generation equipment and the number of hours of power generation, to replace the number of hours of fossil energy power generation, to consume unstable intermittent new energy power generation, to conserve energy, to reduce pollutant gas emissions, and to reduce carbon emissions;
2. The increasing proportion of electric energy storage devices with bidirectional charging and discharging functions, causing increased load fluctuations in the power grid;

3. The problem of irregularity in the behavior of electricity consumption by users and rapid growth and large changes in intermittent loads;
4. The problem of over-expected changes in power loads under the influence of extreme meteorological conditions;
5. the requirement to improve the equipment operation level of distributed resources within the virtual power plant, to realize predictive maintenance, whole-life operable time management, to improve the intelligent operation and maintenance level of power generation equipment and energy storage devices, and the economic operation level of power generation equipment and energy storage devices.

5.5 Commercial Conditions

5.5.1 Brief

The commercial viability of VPPs hinges on three core factors:

resource aggregation capability, which includes adjustment speed and scale;
market rules governing capacity compensation and pricing mechanisms;
and policy support such as priority clearing and subsidies.

As electricity markets mature, VPPs are poised to capture higher revenues in frequency regulation and reserve markets, while optimized load categorization—such as integrating electric vehicle (EV) vehicle-to-grid (V2G) systems—will further enhance their economic efficiency.

5.5.2 Business Models

Virtual power plants (VPPs) operate within diverse business models tailored to market dynamics and operational goals. Direct market participation allows VPPs to bid independently, submitting quantity and price offers to compete in energy markets, while the aggregator model involves collaboration with load aggregators or third-party service providers to streamline participation. The contract model, on the other hand, relies on long-term agreements with grid operators or large consumers, encompassing arrangements like capacity leasing and demand response contracts to secure steady engagement.

5.5.3 Capacity-Price Mechanisms

Revenue generation for VPPs is shaped by capacity-price mechanisms, which blend capacity contributions and market pricing structures. In the capacity-based mechanism, VPPs are compensated based on the amount of capacity they generate.

This mechanism encompasses the peak shaving market, where compensation is tied to adjustable capacity (measured in MW). This market features approaches such as quantity-only bidding—an early-stage policy tool where VPPs declare their adjustable capacity and prices are cleared uniformly—and capacity auctions, where VPPs bid for peak shaving capacity in advance, with payments based on the awarded amount.

In addition to the peak shaving market, the reserve capacity market is another key component, providing payments for committed reserve capacity (e.g., spinning reserves), with energy fees applied only when the capacity is activated. In the reserve capacity market, VPPs bid for the provision of committed reserve capacity, such as spinning reserves, and are compensated based on the amount of reserve capacity they commit. The reserve capacity market is characterized by a dynamic pricing structure, where VPPs are subject to energy market prices, which can be categorized into two main types:

1. **Fixed-Price Mechanisms**: VPPs are directly compensated for the provision of reserve capacity, typically at a fixed rate per MWh of reserve capacity.
2. **Variable-Price Mechanisms**: VPPs are compensated based on the marginal energy price, reflecting the cost of delivering reserve capacity.

Price-Based Mechanisms

Price-based mechanisms further diversify revenue streams, starting with the energy market (electricity trading). In the price-based mechanism, VPPs are subject to energy market prices, which can be categorized into two main types.

Here, peak-valley arbitrage enables VPPs to store energy during low-price periods and discharge it during high-price periods, capitalizing on price fluctuations. Real-time bidding, involving participation in day-ahead or real-time markets with settlement at marginal prices, adds flexibility to energy trading strategies.

The ancillary services market offers additional opportunities, including frequency regulation—where payments are based on performance metrics like response speed and accuracy, often commanding higher prices than the energy market—and demand response, which provides subsidies tied to load reduction (in MWh or MW).

5.6 Technical Conditions

5.6.1 Technical Requirements

The technical level and technical capability of equipment and devices meet the following requirements:

1. Requirements for the level of intelligent automation and digitalization; and
2. Meet the requirements of user experience as well as the requirements of protecting user privacy rights; and
3. Requirements to ensure strong, safe and reliable power system as well as to ensure information security of digital property.

5.6.2 Safe Access and Security

Safe access to power generation equipment, safe access to new hydrogen energy generation equipment, safe access to new energy storage equipment, and technical means to realize fire prevention, explosion-proof, chemical pollution prevention.

5.6.2.1 Automation and Intelligent Monitoring

Realization of automation, intelligent monitoring of distributed energy, intelligent analysis, efficient aggregation, optimization of calculation, economic scheduling, and unmanned automation technology.

5.6.3 Data Visualization

Data visualization, three-dimensional auxiliary design, virtual reality augmented reality display and other good user interaction technologies to improve operation and maintenance efficiency.

5.6.4 Realtime Data

Realize the demand for credible transmission of real-time data on power generation and user load, free from unauthorized use, malicious access and tampering.

References

1. China Huadian Engineering Group Co., Ltd. (Ed.). (2025). *Clean energy and green computing synergy*. China Electric Power Press.
2. Zhong, Y. J., Ji, L., Li, J. X., et al. (2022). Overview of basic characteristics, connotations and development status of virtual power plants. *Integrated Smart Energy, 44*(06), 25–36.

Chapter 6
Planning

Abstract VPPs are special power plants aggregating distributed energy resources to participate in power trading and grid operation. Their operation features diversified resource combination, controllability, openness, and interactivity. Planning involves determining size, architecture, and technical parameters, starting with demand analysis considering external policies and internal grid and user needs. Key parameters are related to clean energy, 5G technology, energy facilities, platform construction, and data management, bringing economic benefits. Phased construction scale and component classification (distributed power, controllable loads, energy storage) are specified. The construction scale should be determined considering various resource aspects, with initial access capacity and adjustable capacity requirements. There are different control topologies (centralized, distributed, fully distributed) with varying characteristics. VPPs also face challenges in technology, market, policy, and operation, with corresponding solutions proposed for better development. They aim to reduce energy costs, enhance efficiency, and meet environmental goals while integrating diverse energy sources.

6.1 General

The planning of virtual power plant is the basis of designing virtual power plant in system configuration scheme, parameter setting, components selection, and final system integration. Planning of virtual power plant includes both the construction of software and hardware that supporting the operation of virtual power plant, and delves into the detailed configuration of each component, providing technical specifications, design considerations, and implementation guidelines.

There is no fixed model in planning of virtual power plant. Since the virtual power plant is a complex system, it needs to adapt to different diversified power resources, for different types of power trading platform to meet all kinds of grid management requirements.

Considerations on how to plan a virtual power plant are elaborated in this chapter. The trade-offs and challenges associated with each option are discussed.

F. Yao, *Engineering Guide to Planning and Design of Virtual Power Plant*,
https://doi.org/10.1007/978-981-95-2231-6_6

6.2 Description of the Planning

6.2.1 Sizing and Performance Definition

Planning of virtual power plant starts in determining the initial size of the virtual power plant and the phased construction and development planning. Then determines the appropriate overall architecture of the virtual power plant platform and parameters defines the performance of virtual power plant.

6.2.2 Optimizing and Smart-Sensing

The planning of virtual power plants should be based on virtual power plants as an energy digitalization platform that coordinates the participation of distributed resources in electricity trading markets and demand response, providing effective measures to address issues related to electricity development. It is advisable to guide users' electricity consumption behavior while meeting their reasonable electricity needs, in line with the goal of comprehensive optimization.

It is advisable to use smart meters and data collection platforms to monitor and analyze users' energy-consuming equipment in real time, understand the load characteristics of the equipment, and combine this with information such as electricity prices to provide users with reasonable energy consumption recommendations.

6.2.3 Demand Analysis

The planning of a virtual power plant project first requires a demand analysis to identify key issues that need to be addressed and requirements that need to be met. Virtual power plants are one of the typical digitalization technologies applied to the energy industry, so the construction of a virtual power plant platform will have the characteristics of openness. The planning of a virtual power plant needs to have the characteristics of openness and scalability, which are also key issues that need to be considered in the demand analysis.

Demand analysis is conducted from two perspectives: external requirements and internal demands. External requirements primarily include regulatory constraints, market mechanism operating rules, and scientific and technological viability. Internal demand factors encompass the balanced and stable power generation of regional grids, the integration of intermittent renewable energy generation, the power consumption characteristics of user loads, the temporal characteristics of user loads, and the construction progress and conditions of supporting power equipment for local grids. Demand analysis aims to simultaneously satisfy both external and internal requirements, achieving the most competitive pricing mechanism while adhering to all constraints.

Specific external requirements to be considered include national energy development strategies, regional grid construction and smart grid regulations. Internal demands commonly include grid peak shaving and frequency regulation requirements, capacity constrains of transmission and distribution facilities in operation, limited conditions in expanding power facilities, uneven distribution of local load over time, and rapid growth of intermittent loads, increased load flexibility requirements of user loads, and heightened grid load fluctuations.

Adopting a virtual power plant solution is an effective approach to addressing the aforementioned internal and external requirements. It is a viable solution that encompasses all elements of a virtual power plant, encompassing power generation, distribution, and consumption.

It enables system peak shaving and frequency regulation, peak shaving and valley filling applications, automated closed-loop control of flexible power, integration of flexible resources, enhanced grid stability, improved power quality, energy consumption, and serving environmental protection issues.

To meet the aforementioned internal and external requirements, it is essential to maintain economic viability and achieve optimal economic benefits.

6.2.4 Benefits

A virtual power plant is a typical example of digital applications in the energy industry. By monitoring, analyzing, aggregating, optimizing, and controlling distributed energy sources (distributed generation, controllable loads, and energy storage). It transforms these into a unified power generation system that can be centrally modulated.

This system participates alongside traditional power plants in system supply-demand balancing and power market transactions, thereby creating new business models and energy service offerings.

By leveraging the adjustable characteristics of virtual power plants as "power generation systems," fluctuations in renewable energy output can be tracked, thereby promoting the consumption of clean energy, improving resource utilization, meeting grid operation and environmental protection requirements, and enhancing overall energy efficiency.

As the power market continues to develop, future power plant generation capacity will depend on strategies for participating in power market transactions. However, single distributed photovoltaic and wind power systems, due to factors such as uncontrollable output and limited capacity, do not meet the access requirements of the power market. Virtual power plants can aggregate and optimize the regulation of distributed energy sources, enabling them to meet the eligibility criteria for participating in power market transactions as a unified entity and thereby generate additional power generation revenue.

By establishing virtual power plants, user electricity consumption behavior can be guided. In conjunction with local power demand-side management regulations, demand-side management can be implemented for users, fully utilizing demand-side load resources to achieve optimized management of comprehensive energy systems and secure specialized subsidies for demand-side response. For example, most steel industries have a certain proportion of production equipment that can be interrupted, such as the electricity-consuming equipment in the steel rolling process, which can reduce or even interrupt electricity consumption. Relevant data indicates that the proportion of load that steel enterprises can reduce is typically around 10.

At the same time, through smart meters and data collection platforms, real-time monitoring and analysis of users' energy-consuming equipment can be conducted to understand the load characteristics of the equipment. Combined with information such as electricity prices, reasonable energy-saving suggestions can be provided to users. Considering load types and development plans, assist users in optimizing energy efficiency and provide efficient electricity management services. Additionally, with the deepening development of power market reforms, users can participate in power market transactions and power ancillary service markets through virtual power plants, achieving a win-win situation for users, the grid, and virtual power plant operations.

6.2.5 Parameters and Specifications

The planning stage should determine the main technical parameters and technical requirements, determine the technical parameters and technical requirements of the hardware platform, and determine the functional composition and technical requirements of the software platform.

6.2.6 System Configuration

The system configuration in a Virtual Power Plant (VPP) refers to the structured arrangement of hardware, software, communication networks, and control strategies that enable the aggregation, coordination, and optimization of distributed energy resources (DERs). It defines how different components interact to ensure efficient grid integration, market participation, and stability [1].

6.3 Main Parameters and Specifications

Main parameters and specifications includes,

Project Scale

Project scale defines the installed capacity (in MW) of the various energy sources and adjustable loads aggregated within the VPP, serving as a foundational metric for understanding its operational scope and resource diversity. Key components include:

- Wind Power: The total installed capacity of wind turbines integrated into the VPP, reflecting the contribution of wind energy to the aggregated portfolio.
- Photovoltaic: The combined capacity of solar PV systems, capturing the potential of solar energy generation within the VPP.
- Controllable Load: The aggregate capacity of demand-side loads (such as industrial processes or commercial HVAC systems) that can be flexibly adjusted—either reduced or shifted—to help balance grid supply and demand.
- Energy Storage: The total capacity of battery storage systems, which store excess energy for later use, enhancing the VPP's ability to manage fluctuations in supply and demand.
- Electric Vehicle Charging Stations: The maximum power capacity of EV charging infrastructure, accounting for both the energy drawn by EVs and their potential to feed energy back to the grid (V2G) in some configurations.
- Other distributed resources: The combined capacity of small-scale, decentralized energy sources like gas-fired micro-generators, biomass facilities, and other renewables (e.g., small hydro), adding further diversity to the VPP's resource mix.

Technical Performance Indicators

Technical performance indicators describe the VPP's operational capabilities, quantifying its efficiency, flexibility, and reliability in grid integration. These indicators include:

- Generation Capacity: Total power output capacity of the aggregated resources, representing the maximum power the VPP can deliver at any given time.
- Annual Generation: Expected yearly energy production (in MWh) from the VPP's combined energy sources, reflecting long-term output potential.
- Regulation Capacity: The ability to adjust power output or demand in response to grid signals, critical for maintaining supply-demand balance.
- Response Time (Large-Scale Adjustments): Time required to implement significant operational changes, such as scaling up/down aggregated resources for prolonged grid needs.
- Response Time (Real-Time Needs): Fast response capability to address immediate grid requirements, such as sudden frequency or voltage fluctuations.

- Regulation Rate: The speed at which the VPP can ramp its output up or down (measured in MW per minute), indicating agility in adapting to dynamic grid conditions.
- Regulation Deviation Rate: The acceptable margin of error in power adjustments, ensuring operational precision and compliance with grid standards.

Grid Connection Operation Indicators

Grid connection operation indicators ensure the VPP's reliability, security, and compliance with grid standards, focusing on the performance of integration infrastructure and monitoring systems. These indicators include:

- Aggregation Equipment Remote Signal Correct Operation Rate: The percentage of remote control signals that function accurately, ensuring reliable command execution for aggregated assets.
- Monthly Availability Rate of Security Protection Equipment: The operational uptime of cybersecurity tools (e.g., firewalls, intrusion detection systems) per month, ensuring compliance with data protection and grid security protocols.
- Remote Measurement Accuracy Error Rate: The allowable deviation between measured and actual values in remote monitoring, reflecting the precision of real-time data used for grid management.
- Monthly Operational Qualification Rate of Electricity Measurement Devices: The percentage of metering equipment that meets regulatory standards monthly, ensuring accurate billing and energy transaction records.

6.4 Phased Construction Scale of the Virtual Power Plant

6.4.1 Scale

The planning of the virtual power plant shall include the construction scale of distributed energy resources, as well as the functions and composition of the virtual power plant hardware platform and software platform accordingly.

6.4.2 Openness

The development scale of distributed energy resources in the virtual power plant shall follow the mode of open access and dynamic development, and it is appropriate to plan in stages according to the initial construction scale, the stage development scale, and the final development scale.

The openness of a Virtual Power Plant (VPP) refers to its ability to integrate new technologies, expand functionalities, and adapt to evolving grid requirements over time. In phased planning, openness ensures that the VPP can scale efficiently, incorporate future DERs, and comply with regulatory changes without requiring a complete system overhaul.

6.4.3 Scalable Resources

The construction scale of distributed energy resources in the virtual power plant shall be determined in accordance with the categorization of power-side resources, load-side resources and energy storage-side resources, such as aggregated resource capacity, adjustable capacity and adjustable response time.

6.4.4 Adjust Capacity

The adjustable capacity at different stages of virtual power plant planning should not be less than 10% of the total load of the virtual power plant.

Increase in power generation of power supply side resources is counted as positive capacity, and decrease in power generation is counted as negative capacity of the same kind.

Decrease in controllable load capacity of load side resources is counted as positive capacity of adjustable capacity and increase in controllable load is counted as negative capacity of adjustable capacity.

6.4.5 Response Time

The response time and duration of Adjustable Capacity's continuous participation in power dispatch shall be summarized in categories of day-ahead generation schedules (one or more days in the future), intraday generation schedules (1 h to several hours in the future), and real-time generation schedules (5 or 15 min to 1 h in the future).

6.4.6 Phased Software and Hardware Platform

The hardware and software platforms for the virtual power plant should be built in one go in accordance with the final scale of construction. And the possibility of phased expansion is reserved. Intelligent terminal with the increase of the construction scale, the intelligent terminal shall be able to realize the control and management of distributed energy resources in the virtual power plant.

6.5 Resources in the Virtual Power Plant

6.5.1 Software Platform

The planning of the virtual power plant should include a variety of distributed energy resources that have access to the power grid, including distributed power sources, controllable loads, and energy storage devices, etc., and the above distributed energy resources should meet the requirements of the grid system trend and stability requirements [1, 2].

Multiple distributed energy resources connected to the grid are shown in Fig. 6.1.

6.5.2 Distributed Power Source

The distributed energy sources included in the virtual power plant shall be distributed power generation units that are close to the power loads, with an installed capacity of several kilowatts to tens of megawatts, and that meet environmental requirements. Priority shall be given to clean energy power generation devices, including gas turbines, internal combustion engines, photovoltaic power stations, fuel cells, wind power stations, geothermal energy power stations, new hydrogen power stations, tidal energy power stations, etc.

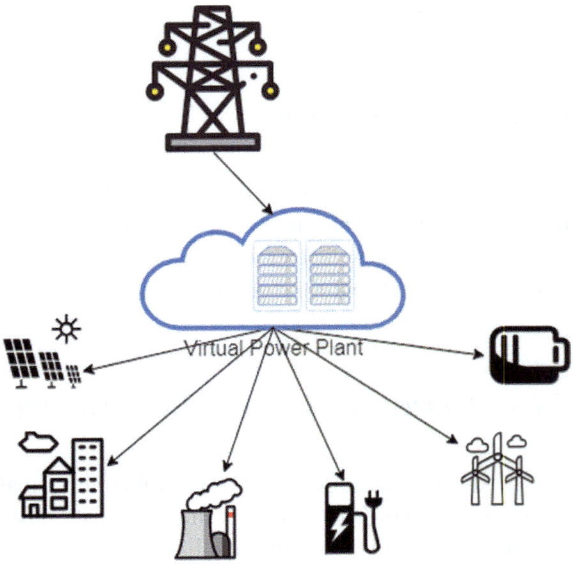

Fig. 6.1 Multiple distributed energy resources access to the power grid

6.5.3 Controllable Loads

Controllable loads included in the virtual power plant shall be loads that can be adjusted in real time according to energy demand under the control of the virtual power plant. New energy electric vehicles in charging mode and demand response loads with in-home controllers can be categorized as controllable loads as well.

6.5.4 Energy Storage Devices

The energy storage devices included in the virtual power plant shall be those that can convert electrical energy into other forms of energy, store energy when there is sufficient power, and can be released in the form of electrical energy when needed by the grid. This includes distributed energy storage and centralized energy storage. New energy electric vehicles capable of maintaining a bi-directional charging and discharging state shall be categorized as distributed energy storage devices.

6.5.5 Control Structure

The planning of the virtual power plant shall be based on a suitable control architecture that integrates the included distributed energy resources under the management and control of the control center of the virtual power plant by means of network communication. The control center of the virtual power plant should include a combination of core software systems and hardware resources of the virtual power plant. Under different control architectures, the control center realizes similar intelligent control functions for distributed energy resources.

6.5.6 Hardware Platform

The hardware platform of the virtual power plant shall include hardwares at least for a control center, a communication system, and decentralized edge smart terminals such as metering units. The resources of the virtual power plant, which are distributed in different locations, should be connected to the control center through smart terminals such as smart metering units tie with the communication system.

6.5.7 Software Platform

The software platform of the virtual power plant shall run on a hardware platform located in the control center that supports a wide range of intelligent core functions.

6.5.8 Types of Aggregated Resources

A virtual power plant is a power coordination and management system that realizes the aggregation and coordinated optimization of distributed power generation devices, energy storage systems, controllable loads, electric vehicles and other resources through advanced information and communication technologies and software systems, in order to participate in the power market and grid operation as a special power plant. The core of the virtual power plant concept can be summarized as "communication" and "aggregation". The key technologies of the virtual power plant include coordinated control technology, smart metering technology and information and communication technology. The most attractive function of the virtual power plant is the ability to aggregate multiple resources to participate in the operation of the electricity market and ancillary services market, and to provide management and ancillary services for the distribution and transmission grids.

Types of aggregated resources include distributed generation resources, energy storage resources, controllable loads and other resources as listed in Appendix A.

For distributed generation resources

- Solar photovoltaic power generation
- Wind power generation
- Biomass power generation
- Small hydropower generation

For energy storage resources

- Electrochemical energy storage
- Lead-acid battery energy storage
- Fluid flow battery energy storage
- Compressed air energy storage

For Controllable load resources

- Industrial loads
- Commercial loads

For Electric vehicle resources

- Electric vehicle charging resources
- Electric vehicle battery resources

For other resources

- Geothermal power generation
- New hydrogen power generation
- Tidal energy power generation
- Other resources

6.6 Control Topology of the Virtual Power Plant

After completing the demand analysis and the determination of internal and external constraints, it can start to enter the process of concrete implementation of virtual power plant. The initial scale of the virtual power plant is determined according to the specific construction conditions and phased scale [1, 2]. Considering the expansion mode and development plan of the virtual power plant, it is necessary to choose the appropriate overall architecture of the virtual power plant platform.

According to the research and application of virtual power plant at home and abroad, the overall architecture of the virtual power plant platform is selected from the centralized, distributed and fully distributed architectures. The main consideration for the selection of control architecture in the design stage of virtual power plant is the control topology of the control center for the aggregated resources.

6.6.1 Centralized Topology

The centralized topology supports centralized regulation and control, i.e., the virtual power plant platform operator acts as a control center and directly develops power generation and power load consumption and energy optimization and management strategies for distributed energy resources based on the participating power market conditions and grid operation security requirements. This type of regulation and control is suitable for small and medium-sized virtual power plant architecture. The virtual power plant performs direct data collection and regulation of various distributed resources, good in real-time data collection, timely regulation and control, and high accuracy of equipment status with only one level of communication channel.

However, such a model has high performance requirements for the control center, requiring high performance and high reliability, and the scalability of the resources that can be accessed to the virtual power plant is limited by the control center's data processing capability. This topology has the best integration with the existing grid management, usually the virtual power plant and the resources covered overlap with the physical area covered by a single grid management, and grid scheduling is easy to realize.

The centralized topology is shown in Fig. 6.2.

6.6.2 Distributed Topology

Distributed topology supports layer control, i.e., based on multi-agent technology, the virtual power plant platform is divided into multiple layers, and the power generation and consumption units in the lowest layer (e.g., the lowest voltage level) can be locally controlled by the sub-virtual power plants/sub-stations (Virtual Power Plants 1, 2, and 3 in the figure below) to ensure the stable operation of the power grid, and the information on their generation and consumption is uniformly fed back to the control center of the virtual power plant platform in the upper layer.

The virtual power plant platform control center is responsible for the regulation and control of the distributed energy resources of the virtual power plant. The virtual power plant platform control center decomposes and allocates the demand to each sub-virtual power plant according to the grid scheduling and power market demand, and then the sub-virtual power plant formulates the operation scheme for the power generation and consumption equipment associated with it. The power generation and consumption units in the lowest layer (e.g., the lowest voltage level) can be locally controlled by the sub-virtual power plants/sub-stations (Virtual Power Plants 1, 2, and 3 in the figure below) to ensure the stable operation of the power grid, and then the sub-virtual power plant formulates the operation scheme for the power generation and consumption equipment associated with it.

In this way, all levels of sub-virtual power plants/sub-stations have autonomy, which can ensure the real-time data and control reliability. While the system is

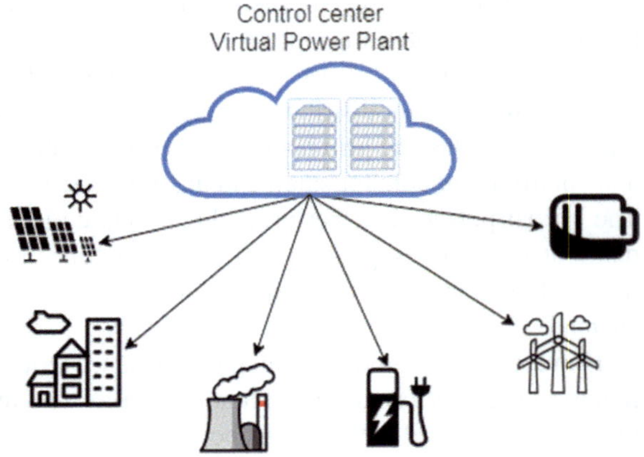

Fig. 6.2 Centralized topology

highly scalable, and the resources that can be accessed are multiplied, and the failure of part of the virtual power plant does not affect the operation of the whole system, thus improves the reliability greatly. The virtual power plants can be backed up by each other, and the resources that can be accessed can be changed at any time by changing the access point of the communication, even in the process of operation.

Even during operation, the accessed resources can be changed at any time by changing the communication access point, which achieves the balance of reliability and scalability. The virtual power plant control center resides in the center and independently performs functions such as data collection, analysis, optimization, and participation in power market transactions, which requires high capacity for big data processing and high requirements for algorithm decomposition and optimization.

This topology is suitable for large-scale virtual power plants. The distributed topology is shown in Fig. 6.3.

6.6.3 *Fully Distributed Topology*

A fully distributed topology supports distributed control, i.e., a fully free agent model where the virtual power plant platform is divided into autonomous subsystems that communicate with each other (Virtual Power Plants 1, 2, and 3 in the figure below), and where each subsystem achieves the tasks that would otherwise be accomplished by the control center through the collaboration of its agents, and where the control

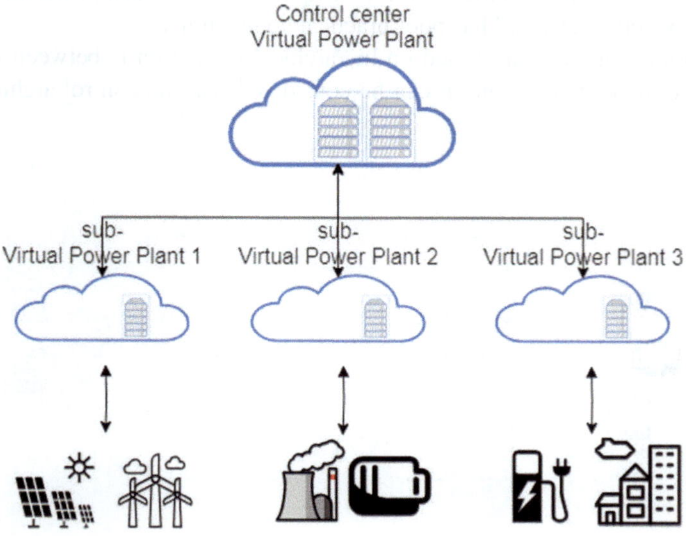

Fig. 6.3 Distributed topology

center is replaced by a data exchange and processing center. This topology of the virtual power plant is suitable for very large, cross-regional virtual power plants.

Each virtual power plant has independent functions for data collection and performance optimization, and does not depend on other virtual power plants or on the data exchange and processing center, but does not participate in the electricity market independently compared to the distributed topology. The virtual power plant data exchange and processing center focuses on data recording, data processing and algorithm optimization, which is able to collect richer big data and better data optimization performance.

The fully distributed topology is shown in Fig. 6.4.

6.6.4 Comparison of Three Network Topologies

Comprehensive comparison of the three network topologies, the control architecture based on centralized regulation and control often requires the control center to have the ability to process large amounts of data and be able to communicate with a large number of end-users, which requires a very high level of data processing and system communication for the control center. The control architecture based on distributed control does not need a control center, but has high requirements for terminal equipment, which is required to be able to collect or receive system operation information and optimize the end-user's energy use process, and therefore requires terminal equipment to have strong data collection, processing and optimization calculation capabilities, and intelligent terminal equipment with such powerful functions is currently very costly and has poor practical applicability.

The control architecture based on hierarchical regulation is between the above two control architectures, and it can be considered that the control architecture of

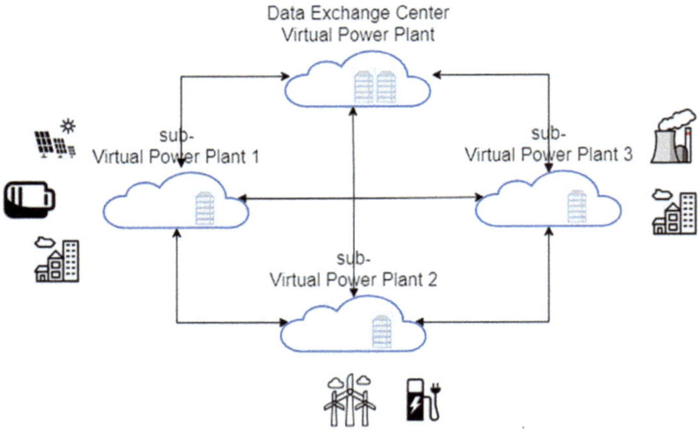

Fig. 6.4 Fully distributed topology

hierarchical regulation is developed on the control architecture based on centralized regulation, and after the gradual formation of a number of virtual power plants based on the control architecture of centralized regulation, the control center of the upper level is added on this basis, so as to form the virtual power plant of the control architecture of hierarchical regulation.

When designing and constructing the virtual power plant control architecture, it is necessary to comprehensively consider the resources of source, load and storage, as well as the operational requirements of the grid and the power market, and carry out detailed research and demonstration comparisons by combining the construction planning of the project, the scale of the resources actually accessed, the geographic location, the demand of the accessing users, the comprehensive performance, the comprehensive investment in the equipment, and the construction revenue.

6.7 Communication Metering and Protection for Virtual Power Plants

In the design of a Virtual Power Plant (VPP), communication, metering, and protection are critical technical components that ensure reliable operation. Below is a detailed analysis of these aspects with some proposed parameter during design.

6.7.1 Communication Design

Communication design for virtual power plants (VPPs) prioritizes three critical considerations to ensure seamless operation: real-time performance, reliability, and data volume management.

Real-time performance varies by function: control commands like automatic generation control (AGC) for frequency regulation demand ultra-low latency—milliseconds to seconds, as specified by standards such as IEC 61850-7-420,[1] which mandates delays under 100 ms to maintain grid stability. In contrast, market transactions are less time-sensitive, tolerating minute-level delays without compromising efficiency.

Reliability is non-negotiable, with communication success rates needing to exceed 99.9% to avoid disruptions. Critical commands, such as those adjusting energy flows, require redundant channels to prevent failures, ensuring continuous operation even if primary links falter.

Data volume depends on the type of information: telemetry from individual distributed energy resources (DERs)—tracking power, voltage, and status—uses 1–10 kbps, manageable via standard networks. However, additional data sources like video

[1] IEC 61850-7-420 Communication networks and systems for power utility automation-Part 7-420: Basic communication structure-Distributed energy resources logical nodes.

Table 6.1 Communication technologies for VPPs

Technology	Use case	Pros and cons
5G/4G LTE	Wide-area coverage, mobility	Low latency (5G URLLC: ~1 ms), but dependent on carrier networks
Fiber optic	Fixed sites (e.g., battery storage)	High bandwidth, low latency, but expensive deployment
LoRa/Wi-SUN	Low-cost, low-power (e.g., rooftop PV)	Low bandwidth (<50 kbps), suitable only for non-real-time monitoring
PLC/HPLC	Existing power line retrofits	No extra wiring, but susceptible to grid noise

surveillance for battery safety monitoring increase demand significantly, requiring 2–5 Mbps, necessitating scalable bandwidth to handle variable loads without bottlenecks.

Together, these considerations balance speed, consistency, and capacity, ensuring VPPs can coordinate DERs, interact with grids, and participate in markets effectively.

VPP communication technology selection uses a hybrid approach, with 5G/4G LTE for wide-area, low-latency needs, fiber optic for fixed high-bandwidth sites, LoRa/Wi-SUN for low-cost, low-power monitoring, and PLC/HPLC for retrofits via existing lines, each balancing pros like latency or cost with cons such as carrier dependence or noise susceptibility (Table 6.1).

6.7.2 Protection System Design

Unique VPP Protection Challenges

Virtual power plants (VPPs) face unique protection challenges stemming from their distributed energy resource (DER) integration.

First, bidirectional power flow—where DERs like solar panels or wind turbines backfeed electricity into the grid—disrupts traditional overcurrent protection systems, which are designed for one-way power flow. This backfeeding can trigger false trips (unnecessary shutdowns) as conventional relays misinterpret reverse currents as faults. To address this, directional relays (such as ANSI 67) are required; these devices detect current direction, distinguishing between normal backfeeding and actual faults to prevent incorrect operations.

Second, high penetration of inverter-based DERs weakens grid conditions by reducing short-circuit capacity—the maximum current a grid can sustain during a fault. This limits the effectiveness of traditional protection systems, which rely on strong short-circuit currents to trigger fault-clearing mechanisms. As a result, adaptive protection settings are necessary: these systems dynamically adjust relay thresholds and operating parameters based on real-time grid conditions, ensuring reliable fault detection and response even as DER output and grid stability fluctuate.

Advanced Protection Techniques

Advanced protection techniques enhance the resilience and responsiveness of virtual power plants (VPPs) amid the complexities of distributed energy resources (DERs).

Wide-Area Protection leverages Phasor Measurement Units (PMUs) to enable ultra-fast, grid-wide coordination. These devices capture real-time voltage and current phasors with high precision, delivering data with sub-100 ms latency. This rapid response allows protection systems to detect and isolate faults across the entire VPP network synchronously, preventing cascading failures and maintaining stability even as power flows fluctuate bidirectionally.

AI Predictive Protection uses Long Short-Term Memory (LSTM) algorithms to forecast potential failures, such as transformer overheating or component degradation, by analyzing historical and real-time operational data. By predicting issues before they occur, this technique enables proactive maintenance, reducing unplanned downtime and optimizing VPP reliability. Unlike reactive protection, it shifts the focus to prevention, aligning with the dynamic and distributed nature of VPPs.

Together, these techniques address the unique challenges of DER-dominated grids, combining speed (via PMUs) and foresight (via AI) to safeguard VPP operations.

A VPP requires communication balancing speed and cost, metering ensuring accuracy and trust, and protection adapting to DER-dominated grids. Its future relies on cloud-edge-device collaboration (e.g., cloud optimization, edge protection, lightweight metering) and AI autonomy.

6.8 Planning of Business Process of the Virtual Power Plant

6.8.1 Business Processes

The diagram illustrates a structured workflow for VPP operations, connecting key stakeholders through coordinated energy trading and grid management activities. The process follows four main phases:

A Virtual Power Plant (VPP) orchestrates distributed energy resources (DERs) like solar panels and batteries into a unified, flexible system. It starts with User Resources-diverse DERs such as rooftop solar or batteries. These feed into the Operation Platform, which first aggregates them. The platform then uses data prediction to forecast energy output and flexibility, with results declared to the Grid Company for market participation.

The Grid Company conducts market clearing, balancing supply and demand across the grid, and determines scheduling data to guide the VPP's resource dispatch. Back at the Operation Platform, this scheduling data drives optimized operation—like charging/discharging batteries or adjusting loads-executed in real time.

The platform also monitors DER performance continuously. This data helps determine assessment metrics (e.g., compliance, efficiency) for financial settlements. For settlements, the platform manages calculations of rewards/penalties and finalizes payments, including compensating DER owners.

The Trading Center organizes energy markets (day-ahead, real-time) where the VPP bids. It logs cleared transactions, calculates fees based on the VPP's performance, and settles financials with all parties. Data flows between layers-from DERs to the platform, grid company, and trading center-ensuring coordination to balance grid stability, market participation, and decarbonization goals, turning scattered DERs into a cohesive, grid-supportive asset [3–5] (Fig. 6.5).

A Virtual Power Plant (VPP) integrates distributed energy resources (DERs) and coordinates multiple stakeholders to function as a unified, flexible energy system with its operational flow as below.

6.8.1.1 User Resources: Foundation of DERs

User-side resources form the building blocks of a VPP. These include:

- **Renewables**: Rooftop solar installations, small-scale wind farms.
- **Storage**: Batteries, electric vehicle (EV) chargers with integrated storage capabilities.
- **Controllable Loads**: Smart thermostats, industrial machinery that can adjust consumption patterns.

Aggregating these diverse DERs gives the VPP the scale and flexibility needed to participate in energy markets and provide grid services.

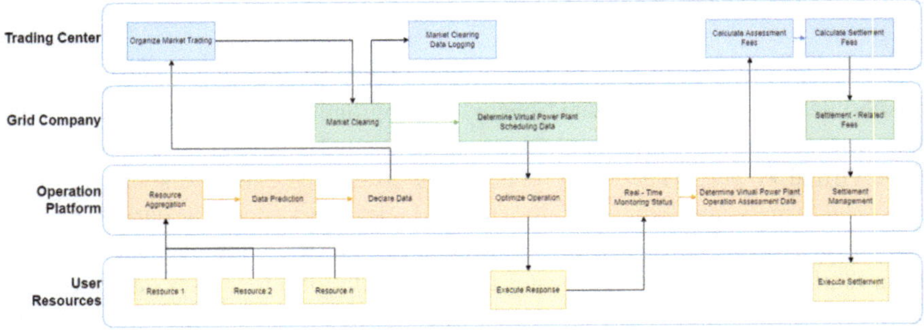

Fig. 6.5 Business process of the virtual power plant

6.8.1.2 Operation Platform: Orchestrating DERs

The operation platform manages daily activities of aggregated resources, balancing technical optimization, market participation, and grid compliance. Key processes include:

Resource Aggregation, Data Prediction, and Declare Data

1. **Resource Aggregation**: Scattered DERs (e.g., 100 residential solar systems + 50 commercial batteries) are combined into a unified virtual "power plant." This creates a scalable pool for market or grid service participation.
2. **Data Prediction**: Leverages historical and real-time data (e.g., weather forecasts, past energy production/consumption) to forecast the VPP's energy output (from renewables) and flexibility (from storage/loads). For example, a machine-learning model might predict 50 MWh of solar production and 20 MWh of battery discharge over 24 h.
3. **Declare Data**: Submits predicted capacity/flexibility to the Grid Company (and indirectly to the Trading Center) to bid in energy markets or secure grid service contracts.

Optimize Operation and Execute Response

1. **Optimize Operation**: Receives "Virtual Power Plant Scheduling Data" from the Grid Company. Uses this to optimize resource dispatch:

   - Prioritize solar/wind output to reduce grid stress.
   - Charge batteries during low-price periods; discharge during high-price periods.
   - Adjust controllable loads (e.g., shift factory operations) to balance supply and demand.

   Example: If the grid needs 10 MWh of peak-time support, the platform might dispatch 8 MWh from batteries and reduce non-critical factory loads by 2 MWh.
2. **Execute Response**: Enforces the optimization plan in real time by sending commands to DERs (e.g., "charge battery now") and monitoring compliance.

Real-Time Monitoring and Assessment Data

1. **Real-Time Monitoring Status**: Continuously tracks VPP performance, including:

   - Energy production (e.g., solar output).
   - Storage status (e.g., battery state of charge, SOC).
   - Load adjustments (e.g., factory consumption changes).
   - Grid conditions (e.g., voltage, frequency).

2. **Determine Assessment Data**: Analyzes monitoring data to generate metrics for:

   - **Compliance**: Did the VPP meet grid service commitments?
   - **Efficiency**: How well did resources perform against predictions?
   - **Reliability**: Did any DERs fail to respond to dispatch commands?

   This data drives financial settlements (rewards for good performance; penalties for underdelivery).

Settlement Management and Execution

1. **Settlement Management**: Calculates financial transactions based on:

   - Market outcomes (e.g., energy sold).
   - Grid service payments/penalties (from assessment data).
   - Costs (e.g., compensating DER owners).

2. **Execute Settlement**: Finalizes payments:

   - Transfers revenue to DER owners (e.g., homeowners with solar panels).
   - Pays the Grid Company for grid services.
   - Settles with the Trading Center for market participation.

6.8.1.3 Grid Company: Balancing Grid Needs

The Grid Company ensures grid stability and integrates DERs into grid operations:

Market Clearing

Validates and processes data from the Operation Platform (and Trading Center) to balance supply and demand. It aggregates bids from multiple VPPs, traditional power plants, and load providers, then decides which resources will:

- Supply energy.
- Provide grid services (e.g., frequency regulation).
- Reduce consumption (demand response).

Example: If there is a solar energy surplus, the Grid Company might clear bids from VPPs to store excess energy in batteries (instead of curbing solar production).

Determine VPP Scheduling Data

Generates a detailed dispatch plan for the VPP, specifying:

- When and how much energy to produce (e.g., "batteries discharge 5 MWh at 3 PM").

- Grid services to provide (e.g., "provide 2 MWh of frequency regulation from 2 to 4 PM").
- Constraints (e.g., "do not exceed 80% battery SOC by 6 PM").

This ensures the VPP aligns with grid needs (e.g., peak shaving, frequency control) while maximizing DER value.

6.8.1.4 Trading Center: Facilitating Energy Markets

The Trading Center creates a marketplace for buying, selling, and trading energy/grid services:

Organize Market Trading

Designs and operates markets where the VPP participates, including:

- **Day-Ahead Markets**: Bid on energy production/sales for the next day.
- **Real-Time Markets**: Adjust bids based on actual grid conditions (e.g., sudden cloud cover reducing solar output).
- **Capacity Markets**: Secure payments for keeping resources available to the grid.

Coordinates with the Grid Company to align market outcomes with grid needs (e.g., ensuring enough capacity for evening peak demand).

Market Clearing Data Logging

Records all cleared market transactions (e.g., "which VPP sold how much energy at what price"). Uses include:

- Audit trails for regulatory compliance.
- Data for market analysis (e.g., identifying price trends, resource performance).

Calculate Fees and Settle Transactions

1. **Calculate Assessment Fees**: Uses VPP operation assessment data to compute rewards/penalties (e.g., bonuses for exceeding grid service commitments; fees for underdelivering energy).
2. **Calculate Settlement Fees**: Aggregates all financial transactions (energy sales revenue, grid service payments/penalties, DER owner compensations) and generates a final "Settlement-Related Fees" report for the Grid Company and Operation Platform.

Grid-friendly virtual power plant aims to aggregate management and optimization control of controllable loads, distributed power sources, energy storage and other resources through advanced information communication and control technologies, and participate in grid operation and market transactions.

6.8.2 System Architecture

The virtual power plant system architecture is shown in the following figure (Fig. 6.6).
Virtual Power Plant Platform: Operating Under the "Unified Management, Distributed Access, and Collaborative Regulation" Model

The virtual power plant (VPP) platform is built on the **power Internet of Things (IoT)**. It combines **smart edge computing** and **perception terminals**, and uses technologies like **terminal-intelligent sensing**, **data-intelligent fusion analysis**, and **AI-assisted decision-making** to function.

The power resources aggregated by the VPP fall into three main categories: **generation, load, and storage**. Each category has unique operational features and key flexibility attributes:

Generation Side

- **Main Resources**: Distributed photovoltaic (PV) systems and other renewable energy generation setups.
- **Key Feature**: The power output is highly intermittent, meaning it fluctuates due to factors like weather (e.g., sunlight availability for PV).

Load Side

- **Main Resources**: Includes air-conditioning loads, industrial loads, and more.
- **Flexibility**:
  1. Industrial loads can be **shifted across different time periods** (e.g., rescheduling factory operations to use power during off-peak hours).
  2. Air-conditioning loads can be **shifted or even interrupted temporarily** (adjusting when ACs run or pausing them briefly to balance grid demand).

Storage Side

- **Main Resources**: Distributed energy storage systems (like batteries) and electric vehicles (EVs).
- **Key Feature**: Offers high flexibility through **adjustable charging and discharging**. This allows for significant load-shifting—storing excess energy when supply is high and releasing it when demand peaks.

Optimal Resource Aggregation Strategy
The VPP figures out the **best way to aggregate resources** by looking at factors such as **energy cost**, how renewable the resources are, and their availability. Effective **resource management and energy optimization** achieved through this strategy help:

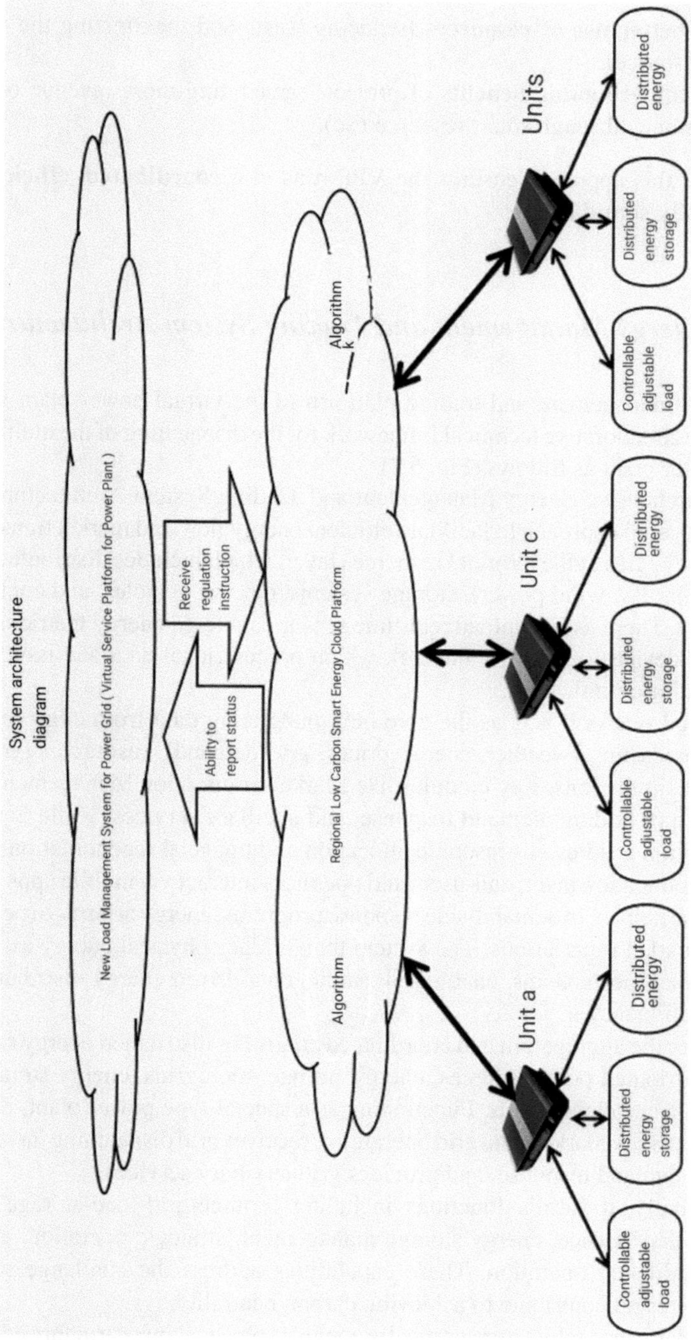

Fig. 6.6 Virtual power plant system architecture

1. **Cut down on energy costs** (by using cheaper energy sources and timing usage smartly).
2. **Make better use of resources** (reducing waste and maximizing the value of each resource).
3. **Boost the economic benefits** of projects (generating more revenue or saving more money through smart resource use).

In short, this approach ensures the VPP runs in a **coordinated, efficient, and economically sound** manner.

6.8.3 Energy Management and Trading System Architecture

The energy management and trading platform of the virtual power plant adopts a cloud-edge collaborative technical framework for the dispatching of the multi-energy virtual power plant as follows (Fig. 6.7).

A comprehensive Energy Management and Trading System Architecture, structured in a layered approach to facilitate efficient energy flow and market transactions.

The flow begins at the terminal resources layer, which includes distributed energy sources (solar PV, wind power), storage systems, electric vehicles, and controllable power users. These assets upload real-time sensor data (e.g., energy metrics) to edge computing devices (gateways, meters), which process local data and issue control commands for optimization.

The cloud platform acts as the core hub, integrating data from edge devices to perform forecasting (weather, energy prices, grid demand), resource aggregation, and market simulations. Key modules like Market Transaction Management handle electricity spot trading, demand response, and ancillary services, while Settlement and Evaluation ensures fair resource allocation and financial reconciliation.

At the application layer, end-users and operators interact via mobile apps or dashboards to participate in demand-side response programs, energy arbitrage (peak shaving), and market transactions. The system thus bridges physical energy assets with digital market mechanisms, enabling dynamic, data-driven energy distribution and economic efficiency.

It enables the aggregation and coordinated control of distributed energy resources such as distributed power sources, energy storage microgrids, energy storage platforms, and controllable loads. Functioning as a special-type power plant, it participates in electricity markets and grid operations, receives grid dispatching instructions, engages in demand response, and provides grid auxiliary services.

Additionally, it fulfills functions including "source-grid-load-storage" power and electricity balance, energy storage management, strategic operation, and optimized coordinated operation. These capabilities address the challenge of power intermittency and contribute to achieving carbon neutrality.

In summary, the architecture vertically connects physical infrastructure edge intelligence cloud analytics market/application services, creating a closed-loop ecosystem for modern energy management and trading.

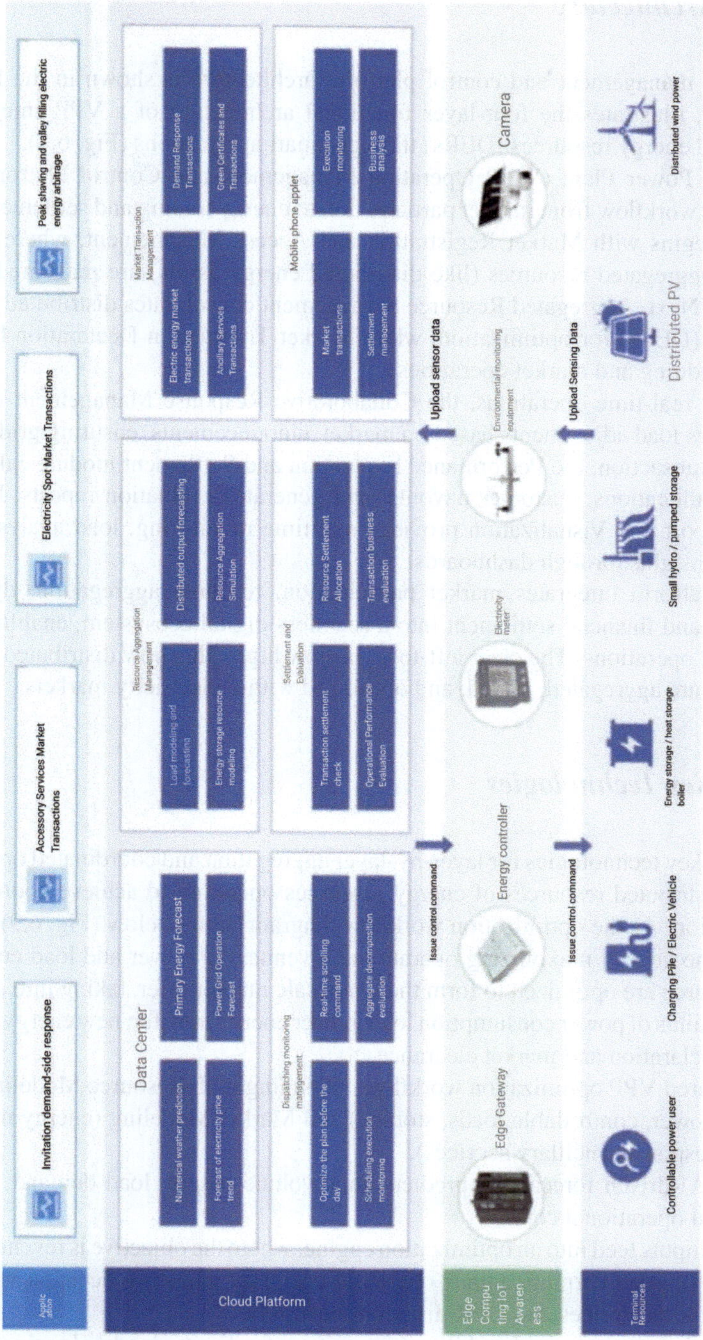

Fig. 6.7 Energy management and trading system architecture

6.8.4 Operation Management and Control Platform Architecture

Operation management and control platform architecture is shown in the following figure, illustrates the four-layer functional architecture of a VPP, integrating distributed energy resources (DERs) through smart applications (Fig. 6.8).

Virtual Power Plant (VPP) Operation Management and Control Platform is a structured workflow from market participation to energy trading and settlement. The process begins with Market Registration and Access Management, where operators and aggregated resources (like distributed energy assets) are registered under contracts. Next, Aggregated Resource Management consolidates distributed energy resources (DERs) for optimization, while Market Transaction Declaration handles pre-day bidding and market operations.

During real-time operations, the Collaborative Response Management module coordinates load adjustments based on market announcements, ensuring grid stability. Post-transaction, the Performance Evaluation and Settlement module calculates resource allocations, customer payouts, and generates evaluation reports. Finally, Data Support and Visualization provides real-time monitoring, load analysis, and business insights through dashboards.

The platform integrates market participation, resource aggregation, dynamic response, and financial settlement into a seamless digital ecosystem, enabling efficient VPP operations. The clear left-to-right flow highlights how distributed energy resources are aggregated, traded, and optimized within electricity markets.

6.8.5 Key Technologies

There are key technologies for layer-by-layer aggregation and coordinated optimization of distributed resources of energy resources operation to achieve coordinated optimization. As the optimization workflow diagram shows below (Fig. 6.9).

With the goal of maximizing operational revenue, the power and load curves of each resource are optimized to form the wholesale market offer, taking into account the constraints of power consumption load, power energy transfer, new energy output, market declaration and market clearance.

Structured VPP optimization workflow, beginning with Resource Modeling (distributed power, controllable loads, storage) and Market Modeling (energy markets, demand response, ancillary services).

Next, AI-driven forecasting predicts photovoltaic output, load demand, market prices, and operational costs.

These inputs feed into an optimization engine, where the objective is revenue maximization (market earnings minus operating costs) by adjusting decision variables like bid volumes/prices and DER dispatch schedules.

Constraints include grid stability (power transfer limits, load balancing), DER operational limits (output/consumption caps), and market rules (bid formatting,

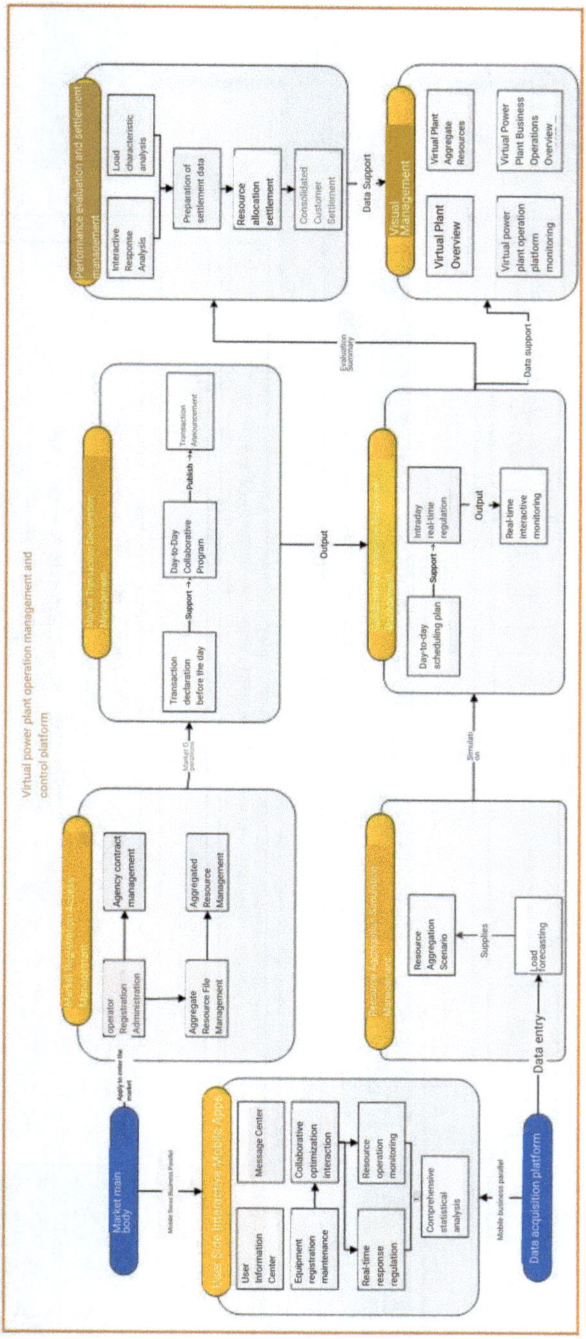

Fig. 6.8 Operation management and control platform architecture

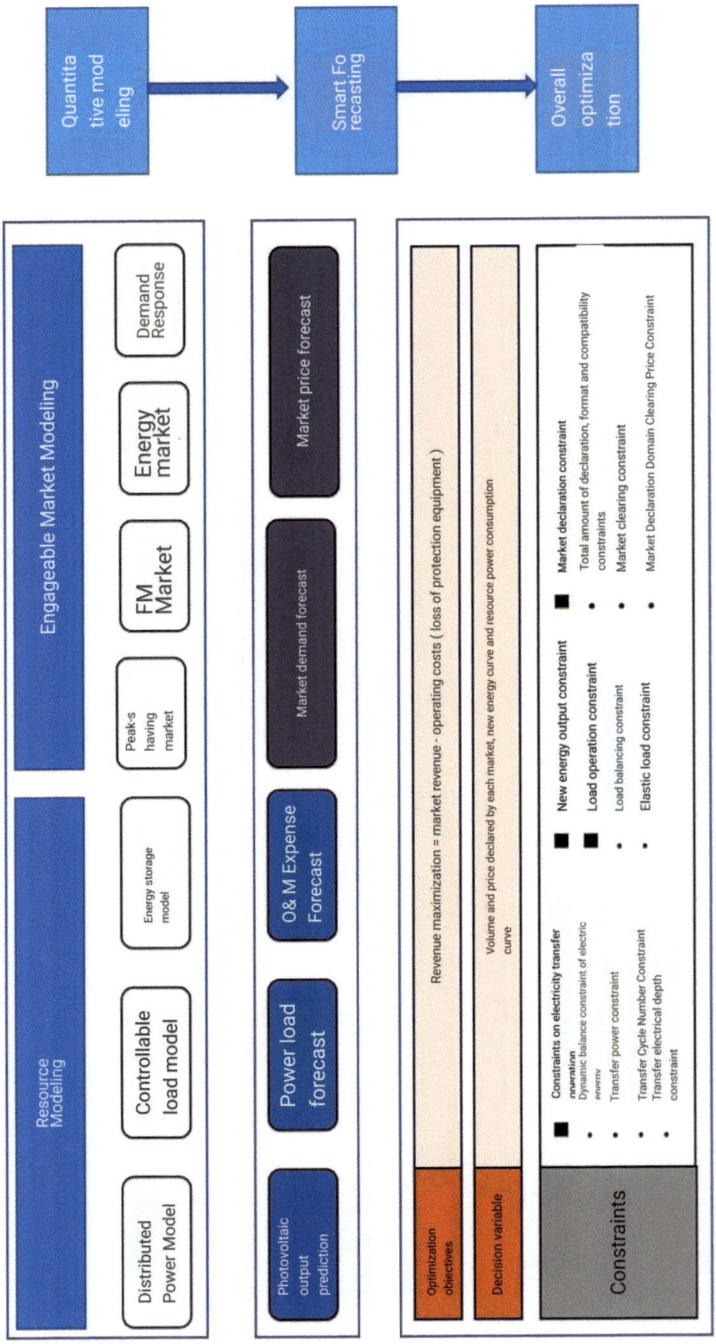

Fig. 6.9 Operation management and control platform architecture

clearing prices). The process culminates in holistic optimization, integrating forecasts, constraints, and market models into a unified strategy for real-time trading and dispatch.

This end-to-end framework enables VPPs to dynamically balance DERs with market opportunities for maximum efficiency and profitability.

The focus of virtual power plant planning and design is the construction of a comprehensive software and hardware platform with a control center, or a data exchange and processing center as the core, and the overall architectural design of the platform construction needs to integrate multiple technologies at different levels, including smart sensing technology based on smart sensors, edge computing and edge control technology, Internet of Things (IoT) technology, big data and cloud computing technology, AI technology, and smart contract technology based on blockchain technology, etc.

The aggregation of all these technologies needs to be tightly focused on electric power energy technology, electric power management system and other professional knowledge, in the form of digitalized information flow, two-way circulation, to achieve intelligent professional management and control.

6.8.6 Economic Benefits

Starting from the market mechanism to analyze the economic benefits of

1. Technical Level

   - Challenge: The virtual power plant involves the integration and synergistic operation of various technologies such as distributed energy, energy storage, intelligent control, information and communication, etc. The complexity of the technology is high, there may be problems such as shortage of technical talents and insufficient technical research and development capability, which will affect the construction and operation effect of the virtual power plant.
   - Solution: Strengthen cooperation with universities and scientific research institutions, introduce professional and technical talents, carry out industry-university-research cooperation projects, and jointly overcome technical problems; actively participate in the industry's technical exchanges, keep abreast of and master the latest technological developments and development trends in the field of virtual power plants, and promote technological innovation and application.

2. Market Level

   - Challenge: The business model and profitability mechanism of the virtual power plant are not yet perfect, and the market rules and price mechanism are not sound, which may lead to market risks and uncertainty of returns when virtual power plant participates in the power market transactions; the users' recognition and acceptance of the virtual power plant is low, and their participation is not highly motivated, which affects the resource integration and market promotion of the virtual power plant.

- Solution: Explore diversified business models, such as providing integrated energy services, participating in the demand-side response market, and carrying out investment and operation of distributed energy projects, to improve the profitability of virtual power plants; strengthen market research and analysis, formulate reasonable market strategies, and provide personalized services and solutions according to market demand and user characteristics.

3. Policy Level

- Challenge: As an emerging power business, virtual power plant is still in the process of improving relevant policies and regulations, and there may be insufficient policy support, lack of policy implementation, etc., which will bring certain policy risks to the construction of virtual power plant.
- Solution: Actively pay attention to the relevant policies and regulations issued by the national and local governments, and strive for policy support and subsidy funds, such as subsidies for renewable energy development, subsidies for the construction of energy storage facilities, and subsidies for demonstration projects of the virtual power plant, etc.; strengthen the communication and coordination with the governmental departments, and reflect the problems and difficulties encountered in the process of the construction of the virtual power plant in a timely manner, so as to push forward the improvement and implementation of the policies.

4. Operation and Management Level

- Challenge: The virtual power plant needs to monitor, dispatch and manage a large number of decentralized distributed energy sources and controllable loads in real time, which makes operation and management more difficult. There may be problems such as insufficient experience in operation and management, low level of management, and imperfect coordinating mechanism, which will affect the operation efficiency and reliability of the virtual power plant.
- Solution: Establish a sound operation and management system for the virtual power plant, formulate scientific and reasonable operation and management systems and processes, strengthen the training and assessment of operation and management personnel, and improve the operation and management level; strengthen the communication and coordination with the power grid enterprises, energy suppliers, users and other parties to establish a good cooperative relationship, and form a synergistic operation synergy.

6.8.7 *Challenges and Solutions*

The construction of virtual power plant faces the following challenges and solutions:

1. Market mechanism: in the initial stage of market trading, virtual power plants and controllable load resources participate in the power peaking market and prioritize thermal power clearing in accordance with the method of quoting quantity but not price, which simplifies the participation process and helps to improve the market participation of virtual power plants. In addition, the rules also stipulate in detail the mode of virtual power plant participation in market transactions, the declaration process, and the rules for clearing, which makes the trading behavior of virtual power plants more standardized and orderly, and reduces the uncertainty and risk in the transactions.
2. Environmental benefits: The development of virtual power plant helps to improve the capacity of renewable energy consumption, reduce the dependence on traditional fossil energy, reduce carbon emissions, realize the sustainable development of energy, in line with the national "dual-carbon" goals and environmental protection requirements.

6.8.8 *Intelligent Control System*

Adopt advanced intelligent control system, use Internet of Things, big data, cloud computing and other technical means to conduct real-time monitoring and data analysis of distributed energy resources, automatically optimize energy production, distribution and storage strategies according to the grid demand, market prices and the enterprise's own energy conditions, realize coordinated control and optimal dispatch of distributed energy, and improve the efficiency of energy utilization and economic benefits.

References

1. Yao, F., Xie, G., Li, C., Liu, X., Ren, B., Lu, H., Li, R., Song, W., Duan, C., & Wang, W. (2024). Flexible construction technology and application of new integrated energy systems. In *Proceedings of the 2024 new power system development forum* (pp. 94–99). China Electric Power Press.
2. Yao, F. (2023). Engineering on the basis of big data in virtual power plant. In *Proceedings of 24th conference on electric power supply industry, Xiamen, China.*
3. IEC TS 63189-2:2023 virtual power plants—part 2: use cases. https://www.en-standard.eu/pd-iec-ts-63189-2-2023-virtual-power-plants-use-cases/
4. Zapata, J., Vandewalle, J., & Dhaeseleer, W. (2014). A comparative study of imbalance reduction strategies for virtual power plant operation. *Applied Thermal Engineering, 71*(2), 847–857.
5. Kaczorowska, D., Rezmer, J., Sikorski, T., et al. (2019). Application of PSO algorithms for VPP operation optimization. *Renewable Energy and Power Quality Journal, 17*(17), 91–96.

Chapter 7
Detail Design and Technical Implementation

Abstract This chapter presents a technical framework for designing virtual power plants (VPPs), critical to digitalized grids and the energy Internet, featuring a layered architecture with four core components: distributed energy resources (DERs), communication networks, data platforms, and intelligent applications, all aligned with industry standards and security protocols. DER design integrates distributed generation, controllable loads, and energy storage, utilizing IoT and AI for real-time digital mapping, forecasting, and optimal scheduling to enhance grid stability and renewable energy adoption, supported by smart sensors and edge devices for monitoring and actuation. The communication layer employs 5G and advanced technologies to establish secure, low-latency bidirectional channels between control centers, DERs, grid platforms, and markets, ensuring reliable data flow. The data platform standardizes data collection, cleansing, aggregation, and visualization (via maps, dashboards) to track resource status, facilitating informed decisions. Intelligent applications drive demand response, market trading, and ancillary services (e.g., frequency regulation) through AI, big data, and machine learning, optimizing forecasting, scheduling, and revenue, with modules for market registration, contracts, settlement, and evaluation enabling seamless participation. A pilot integrating solar PV, batteries, and flexible loads demonstrates stability improvements and cost reduction, leveraging cloud-edge-end IoT, 5G/private networks, and scalable cloud infrastructure. Software integration connects VPPs to external systems, with user interfaces for monitoring and transaction control, emphasizing "software-network-hardware" synergy to unify energy, information, and business flows, advancing renewable integration and grid resilience.

F. Yao, *Engineering Guide to Planning and Design of Virtual Power Plant*,
https://doi.org/10.1007/978-981-95-2231-6_7

7.1 General Provisions

7.1.1 Brief

The design of virtual power plant should be based on the planning of virtual power plant, according to be divided into four components: design of distributed energy resources, design of communication network, design of data hardware platform for database, design of intelligent applications.

7.1.2 Distributed Energy Resources Design

Distributed energy resource design [1] requires the realization of the functional design of devices such as distributed power sources, controllable loads, and energy storage devices, as well as the digital design of the conversion of specific generation and consumption regulation entity operations to data platforms.

Distributed energy resource design builds a flexible and efficient energy network by integrating distributed power supply controllable load and energy storage systems. The core lies in the functional optimization and digital synergy of physical equipment: power supply and energy storage need to match the load demand to achieve peak shaving or frequency regulation; controllable loads dynamically adjust their power consumption behavior through demand response.

Relying on the Internet of Things and cloud platforms, through intelligent data collection terminals, physical power generation/consumption data are mapped to digital systems in real time, and AI algorithms (e.g., load forecasting, optimal scheduling) and virtual power plants and technologies are used to achieve remote monitoring, intelligent decision-making and market-based trading, ultimately enhancing renewable energy consumption and grid stability.

7.1.3 Smart Sensing and Acutation

For the design of a virtual power plant, a variety of sensors and actuators should be installed on various energy resources, enabling capabilities of intelligent perception and edge control.

These sensors can intelligently perceive data like energy generation, consumption, and equipment status. Actuators, in turn, enable edge control, allowing for real-time, localized adjustments. For example, the intelligent perception from sensors and edge control via actuators can feed into the communication network and upper-layer smart applications, facilitating coordinated management of energy resources, from distributed generation to smart dispatching and market transactions.

7.1.4 Communication Network Design

Communication network design [1], realize the design of communication channel and communication equipment, realize the two-way transmission of control information and data information such as multi-level control center, bottom distributed energy resources, etc. in accordance with the needs of control architecture. Communication technologies including 5G shall be adopted to realize ubiquitous connectivity of the energy system.

7.1.5 Data Hardware Platform Design

Data hardware platform design, realize the hardware design of the control center of the virtual power plant and the intelligent terminals of the underlying distributed energy resources, and construct a data platform for the operation of the virtual power plant.

7.1.6 Intelligent Application Design

Intelligent application design, realizing virtual power plant entity business datatization, intelligent terminal metering and control information datatization, communication network layer data security management, realizing intelligent computing, big data computing, distributed energy resources datatization fusion and data optimization, realizing power scheduling, power market trading and other intelligent application layer business processes.

7.1.7 Smart Data Processing and Analysis

Virtual power plant detailed design on data processing, need to be based on the communication network layer in the distributed energy resources layer of energy operation state information and data collected, analysis and mining, the analysis of data research and related knowledge of the energy industry, the results of the data analysis, and the actual power and energy scheduling and management, as well as market transactions.

7.1.8 Software-Network-Hardware Synergistic

The design of the virtual power plant should realize "software-network-hardware" synergistic digital ecological business, digital-based information flow and energy conversion and transmission-based business flow, two-way fast flow, interactive integration. From the grid scheduling side and the power trading market side, the virtual power plant aggregates multiple resources to achieve the equivalent performance and functions of the power plant, and comprehensively meets the various needs to achieve the function of power plant virtualization [2–4].

Here is a architecture shows how Virtual Power Plant (VPP) works (Fig. 7.1).

At the edge, smart terminals connect metering units monitoring adjustable loads, energy storage devices, and distributed generation resources.

A wireless/leased/optical network transmits data to the Virtual Power Plant Control Center, which uses intelligent function modules for real-time DER mapping, forecasting, and scheduling.

The control center interacts upward: with grid dispatching/management for stability services (e.g., frequency regulation) and electricity market transactions for trading, registration, and revenue optimization, integrating energy, info, and business flows to boost renewable uptake and grid resilience.

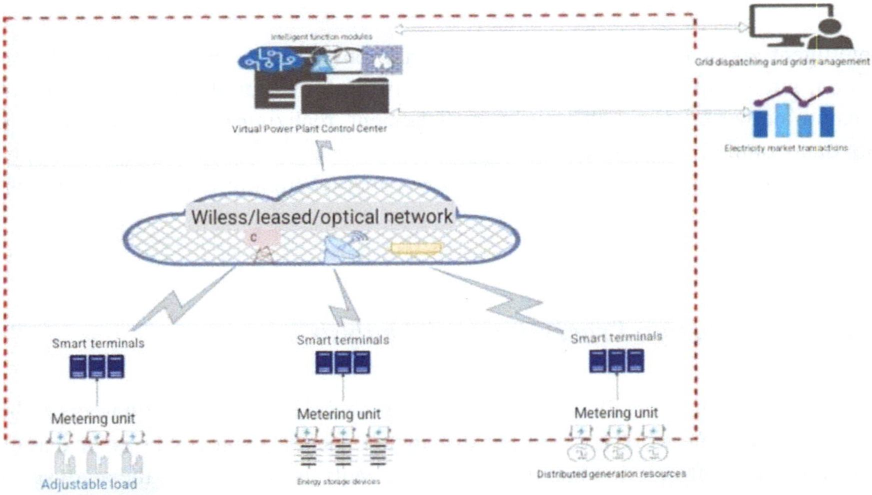

Fig. 7.1 Software-network-hardware synergistic architecture

7.1.9 Data Management

The data of the virtual power plant should strictly comply with the corresponding data management regulations to ensure information security, so that the data can be seen, the flow can be monitored, the risk can be perceived, the leakage can be traced, and the encryption can be used. Based on the requirements of the digital technology application of electric energy, the industry standard regulations and management rules of the electric power system should be complied with, and the industry standard specifications and management rules of the information system should be complied with.

It shall meet the requirements of information security physical protection security, independent and controllable intellectual property security, and software protection security.

7.2 Overall Architecture and Design of Components

Design of virtual power plant should be in line with the requirements of the layered architecture of Appendix A.4 (Fig. 7.2).

According to the layered architecture design of the virtual power plant platform, the specific construction content of the virtual power plant platform should be realized through the construction of the distributed energy resource layer, the construction of hardware platform for the data collection and processing layer, the construction of the aggregation intelligent application function layer and the communication of each layer [5].

Construction of Distributed Energy Resource Layer

The construction of the distributed energy resource layer should realize the function that in order to realize the access of resources such as source, load and storage to the virtual power plant platform, and to realize their participation in the dispatch operation of the electric power system, electric power market transactions and ancillary services on behalf of the virtual power plant platform, it should carry out a certain amount of special transformation of the equipment of the existing resources of the source, load and storage, empower the energy equipment by means of digital technology, realize the access to its communication and remote control, and meet the requirements for the measurement accuracy and sampling of the operation, control and transaction.

Control and trading measurement accuracy and sampling requirements. For example, for large industrial users, such as participating in the demand response platform to guide industrial users to adjust the production of electricity in the form of invitations, it is indirect control without direct access, and only the user's metering terminals need to be transformed to meet the requirements of metering and settlement; while for

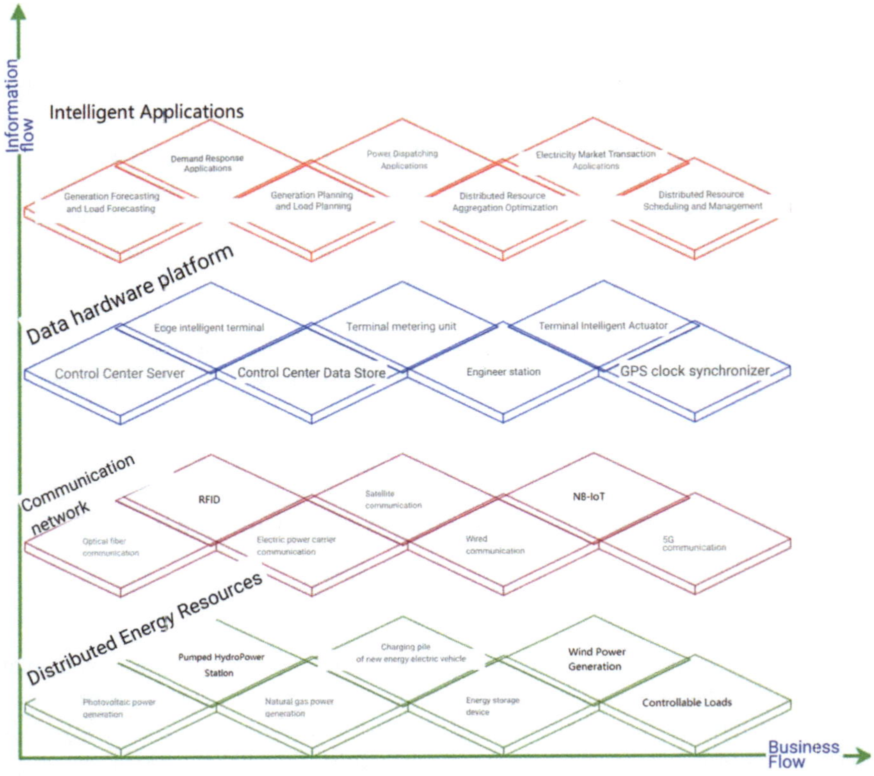

Fig. 7.2 Layered architecture of the virtual power plant

users that can be directly controlled, it is necessary to install intelligent control terminals, which can be multiplexed with fiber-optic channels and other communication modes, to achieve the main station of the virtual power plant platform to carry out remote and direct control. Remote direct control.

Various types of source, load and storage energy resources are transformed according to their actual conditions, for example, the central air conditioner of the building needs to carry out certain communication interface development, so that it can communicate the collection and control information with the installed intelligent terminals, and access the master station of the virtual power plant platform through the 4G/5G public network, etc. For the different characteristics of various types of resources, it is necessary to carry out a feasibility study of the special transformation one by one and adopt the program to implement it.

Construction of Hardware Platform for Data Collection and Processing Layer

The construction of data collection and processing layer should combine the distribution of various types of resources in source, load and storage, and realize the unified collection and access of various types of resource data through standardized data format and standardized data collection interface. Meanwhile, by combining the attribute differences of source, load and storage model data, and considering the characteristics of data type, refresh scale, change range and correlation relationship, the resource data shall be classified and processed, anomalous data identified, data calibrated, data analyzed, and alarm monitored, etc., so as to realize data cleansing and data quality auditing, and to avoid the interference of inappropriate data to the operation of the virtual power plant platform. On the basis of the object attributes of individual energy resources, modeling and aggregation of individual energy resources are carried out from the dimensions of space, time, control attributes, etc., to form the regulation aggregation model with different levels, time and control objects, assess the adjustable range of the virtual power plant, and provide a reference range of regulation for the subsequent participation in the scheduling and operation and the power market transaction. It also realizes visualization to show the basic information of source, load and storage resources, adjustable capacity, real-time output and other key information, so as to realize the monitoring of the operation status of various types of resources.

Construction of Aggregation Intelligent Application Function Layer

The construction of the intelligent application function layer of aggregation should be combined with the relevant policies of the industry, and different forms of services should be established in phases, such as participation in demand response aggregation application, and the aggregation application for participation in ancillary services and power market transactions should be built gradually in the regions with ancillary service conditions.

Construction of Intelligent Communication Network

The construction of the communication channel needs to be based on the actual situation of the source, load and storage energy resources, special communication deployment and access program design, to open up the communication channel between the energy resources, the virtual power plant platform, the scheduling platform, the power market trading platform, etc., so as to realize the application functions of the virtual power plant.

Intergration of Layers

The overall architecture design of the virtual power plant platform is the key link in the implementation of the virtual power plant project, connecting the actual business scenarios and digital business processes.

Under the overall architecture of the virtual power plant, various functions are eventually realized and fully integrated in a high-performance data and information system to realize a digital virtual power plant.

From the perspective of both the grid scheduling side and the power market trading side, a geographically independent virtual power plant is obtained, and the various resources accessed by the virtual power plant, whether in a centralized, distributed or fully distributed architecture, are displayed on a unified platform and optimized for scheduling and control. The design of virtual power plant should realize at least include relevant functions as below.

The integrated operation system of virtual power plant adopts advanced information technology integration to realize automatic measurement and collection of data of distributed power generation, energy storage devices, controllable loads, etc., and real-time collection, statistics, analysis and display of operating indexes such as circuit breaker switching status, voltage, current, active power and reactive power of each resource, energy consumption indexes, so as to realize refined measurement, visualized monitoring and intelligent management.

Data visualization is used to monitor and display instantly by means of geographic area map, pie chart, histogram, heat map, river map and so on.

The parameters of metering and monitoring objects are analyzed on a daily, weekly, monthly, quarterly and yearly basis, and the daily, weekly, monthly, quarterly and yearly energy consumption reports of each node are automatically formed.

7.3 Distributed Energy Resource Layer

7.3.1 Configuration

Distributed energy resource Layer integrates distributed energy resources, and configuration of virtual power plant intelligent terminal should realize the following functions.

The distributed energy resource layer brings together distributed energy resources. For the configuration of the virtual power plant intelligent terminal, it should achieve the following:

First, it needs to collect data from the connected source, load, and energy storage resources.

Second, it must have a communication function to send the gathered data to either the virtual power plant main station control center or the local virtual power plant sub-station control center.

Third, it should be capable of implementing specific local control functions based on the requirements of different user resources.

7.3.2 Access to Resources

Distributed energy resources are designed to realize the access of resources such as power generation, load and storage to the virtual power plant platform, and their participation in power system dispatch operation, and power market transactions and ancillary services is represented by the virtual power plant platform.

7.3.3 Transformation Design

For distributed energy resources design, according to the requirements of the construction conditions, certain transformation design for the existing equipment shall be carried out to realize its communication access and remote control, so as to meet the operation control and transaction of the measurement accuracy and sampling requirements.

7.3.4 Intelligent Control Terminal Design

For users that can be directly controlled by the virtual power plant, intelligent control terminals should be installed, and communication methods such as multiplexed fiber optic channels can be used to realize remote and direct control by the master station of the virtual power plant platform.

7.3.5 Intelligent Metering Terminal Design

For users whose demand response platform guides industrial users to adjust the indirect control of production power consumption in the form of invitation, the metering terminals of the users shall be remodeled to meet the requirements for metering and settlement.

The virtual power plant intelligent terminal is installed at the first level and performs three core functions.

First, it collects data from connected energy resources, including generation sources, loads, and storage systems.

Second, it features communication capabilities to transmit the collected data to the virtual power plant platform's main station or a local sub-station.

Third, it enables localized control based on user requirements, allowing energy resources to independently decide whether to participate in virtual power plant operations at any given time.

7.4 Communication Network Layer

7.4.1 Communication Deployment Scheme

The communication network design should be carried out according to the actual situation of different sources, loads and storage resources and the access requirements of the power grid to carry out the design of the communication deployment scheme, and realize the communication functions among intelligent terminals, the master station and sub-stations, as well as the external communication functions of the virtual power plant master station. The communication to the dispatching platform and the power market trading platform needs to meet the corresponding information security requirements.

7.4.2 Communication Channel

The design of the communication channel should be based on the actual situation of the source, load and storage energy resources, and the design of special communication deployment and access programs should be carried out to connect the communication channel between the energy resources, the virtual power plant platform, the scheduling platform, and the power market trading platform.

7.4.3 Bidirectional Communication

A bidirectional communication channel should be established between the control center of the virtual power plant and the controlled intelligent terminals to ensure fast and smooth data communication from the physical layer and the data link layer.

7.4.4 Reliable Communication

A reliable communication channel shall be established between the control center of the virtual power plant and the outside of the virtual power plant to support the virtual power plant to participate in the power market transaction and accept the scheduling and management of the power grid.

7.4.5 *Cost and Performance Balance*

The communication network design shall adopt advanced and reliable technologies, and on the basis of economic rationality, use industrial internet, 5G communication and other advanced and applicable technologies to ensure smooth data channels and realize data synchronization and time delay compensation.

7.4.6 *Security of Transmitted Data*

Communication network design shall ensure the security of transmitted monitoring data and measurement data.

7.5 Data Hardware Platform Layer

7.5.1 *Function Design*

The design of the data hardware platform involves configuring the virtual power plant data platform to achieve the functions of the virtual power plant master station. It enables the monitoring of the energy status of sources, loads, and storage, including the storage, monitoring, and analysis of collected data.

Additionally, it delivers the calculated and analyzed results of corresponding application functions to smart terminals, thereby completing the regulation of sources, loads, and storage.

7.5.2 *Data Collection and Access*

The data hardware platform design should realize the unified collection and access of various types of resource data through standardized data format and standardized data collection interface.

At the same time, combining the attribute differences of source, load and storage model data, and taking into account the characteristics of data type, refreshment scale, change range and correlation relationship, the resource data shall be classified and processed, anomalous data identified, data calibrated, data analyzed, and alarm monitored, etc., so as to realize data cleansing and data quality auditing, and to avoid the interference of inappropriate data to the operation of the virtual power plant platform.

7.5.3 Data Visualization

The design of data hardware platform should realize visualization, show the basic information of source, load and storage resources, adjustable capacity, real-time output and other key information, and realize the monitoring of the operation status of various types of resources. Adopt data visualization with geographic area map, pie chart, histogram, heat map, river map, etc. for instant monitoring and instant display.

7.5.4 Data Logging

The data hardware platform is designed to process the operation data of each distributed power source, energy storage device and controllable load in the area included in the virtual power plant, and to provide effective methods of saving, recording and obtaining, which can be used in the comprehensive coordination and optimization and decision support system.

It realizes that the parameters of the metering and monitoring objects are analyzed on a daily, weekly, monthly, quarterly and yearly basis, and the daily, weekly, monthly, quarterly and yearly energy consumption reports of each node are automatically formed.

7.5.5 Data Aggregation

The design of data hardware platform should aggregate intelligent terminals of multiple resources to realize automatic data metering and collection of distributed power generation, energy storage devices, controllable loads and other aggregated resources, which can be used for real-time collection, statistics, analysis and display of operating indexes including circuit breaker switching status, voltage, current, active and reactive power, and energy consumption indexes, so as to realize refined metering, visualized monitoring and intelligent management.

7.6 Intelligent Application Layer

7.6.1 Configuration

Intelligent application design, configuration of virtual power plant application functions, realization of optimized regulation of source, load and storage, participation in demand response, power system scheduling and power market transactions.

7.6.2 Demand Response Application

Participation in demand response aggregation applications shall meet the following requirements:

1. Applications participating in demand response shall interact with the marketing demand-side response platform of the regional power supply company in which the virtual power plant is located, and participate in the operation and management of power controllable loads organized by the platform; and
2. The demand response platform, which shall be adapted to the different modes of invitation-based, automatic demand response and market-based operation, shall sign contracts with large users and a small number of low-voltage side users with regulating characteristics, and carry out regulation by invitation; and
3. The demand response platform shall be required to have such functions as registration management, quantity quotation, issuance of regulation instructions, optimization of regulation execution, regulation metering management, settlement management and effect evaluation.

7.6.3 Participation in Ancillary Service Application

Participation in ancillary service application shall meet the following requirements:

1. The function of participating in power market trading application can enable the virtual power plant and other power generation resources to participate in the future trading of electric energy spot market, which will determine the day-ahead, real-time planning and scheduling arrangements and ancillary service clearing quantity in a market-oriented form, and obtain the marginal price of the spot market and the value of each node's electric energy and the marginal value of ancillary services; and
2. Functions participating in the ancillary service aggregation application shall have functional modules such as registration management, user management, transaction management, receiving transaction demand, receiving calibration of clearing information, optimizing regulation execution management, settlement management, and effect evaluation.

7.6.4 Intelligent Function Modules

The virtual power plant shall set up intelligent function modules, and the design of intelligent function modules shall meet the provisions.

Generation Forecast and Load Forecasting The virtual power plant shall forecast the power generation and load, and the intelligent function module shall realize the generation and load forecasting, and the realization of the load balancing of the generation and load forecasting.

Preparation of Generation Plan and Consumption Plan

The virtual power plant shall prepare the generation plan and consumption plan based on the historical operation data, load forecast data, power generation forecast data, and other data, and the intelligent function module shall realize the optimization of the generation plan and consumption plan, and the realization of the load balancing of the generation plan and consumption plan.

Controllable Load Management

The virtual power plant shall manage the controllable load, and the intelligent function module shall realize the optimization of the controllable load, and the realization of the load balancing of the controllable load.

Control and Management of Energy Storage Devices

The virtual power plant shall manage the energy storage devices, and the intelligent function module shall realize the optimization of the energy storage devices, and the realization of the load balancing of the energy storage devices.

Coordination and Optimization of Distributed Generation

The virtual power plant shall coordinate and optimize the distributed generation, and the intelligent function module shall realize the optimization of the distributed generation, and the realization of the load balancing of the distributed generation.

7.6.5 Functional Design

Virtual Power Plant aggregates distributed energy resources (DERs) into a unified, intelligent network. And intelligent module in a VPP enhances decision-making through AI/ML, real-time optimization, and automation. The virtual power plant shall synthesize functions into one unified system for intenlligence modules with the following functions.

7.6.5.1 Marketing Function

Market Registration Access Management

Provide market member full life cycle management functions. Market member management includes information management of virtual power plant operators and market members participating in the transaction, resource access management, resource file management when applied, and provides registration, change, and withdrawal management of members' qualification registration and virtual resources and other information, as well as maintenance of various types of user registration information participating in regional multi-energy transactions. The main functions include file management of market members, agency relationship (contract) management and management of related resources.

Aggregator Registration Management

Provides unified aggregator registration management function, supports the registration application of virtual power plant resources in different zones, and provides the maintenance of basic information of market members and the process management of registration application. Market members fill in the basic information, including the credit code and other qualification information of the enterprise, as well as the enterprise's scale of electricity consumption, enterprise type and other basic information, and send the registration application to the virtual power plant operation and control platform; after the application has been approved, further improve the file information of the virtual power plant, including the enterprise's credit code, the name of the enterprise, and other basic information that can not be modified, and if it is modified, it is necessary to be re-examined and approved; improve the resource information of the enterprise, including the virtual power plant file information, the virtual power plant name, and other basic information. Including virtual power plant file information, virtual power plant virtual power generation resource information, virtual power plant virtual generating unit file information. If the audit is not passed, it will be knocked back and resubmit the registration application.

Resource Access Management

Realize safe access to distributed energy, energy storage, load and other diversified resources and heterogeneous platforms in the park. Supports 5G/4G, LORA, RS485, WIFI, and Ethernet heterogeneous communication. Supports monitoring and analyzing the accessed resources, statistical analysis, and necessary data cleaning and processing.

Agent Contract Management
Contract management for the accessed virtual power plant agent resources, through the contract templates, terms and conditions and electronic, etc., to achieve intelligent

management of user contracts. It provides contract templates, contract signing, contract change, contract execution, contract ledger and other functional points. It mainly provides the functions of formulating, managing and maintaining different types of contracts, and provides the whole process management of formulating, approving, signing, executing, changing, terminating, archiving and inquiring of all kinds of contracts and agreements, as well as the corresponding technical support.

Market Transaction Declaration Management

Market transaction declaration management function provides virtual power plant operators and agents to participate in peak and frequency regulation, emergency power and electric energy transactions in the power market, supports the business management of virtual power plant market-oriented transaction operation before, during and after the transaction, and provides technical support for the transaction process such as program preparation, transaction declaration, transaction results, and the function of plan preparation.

Transaction Announcement Management

Provide management functions of transaction announcement information and transaction constraint information. Provide the maintenance function of transaction results; the transaction results can be viewed by power plant operators.

Transaction Program Management

According to the information of transaction announcement, new transaction declaration program can be added, and can be modified and deleted before the program is audited; the program cannot be adjusted after the program is audited; the program includes the load curve, time period, and resource declaration plan; when editing the transaction declaration program, the distributed resource plan can be added or deleted according to the needs of the program.

Market Declaration Management

On the basis of load forecast and price forecast, combined with grid regulation demand and operator declaration, it supports different aggregation units of virtual power plants to make market declaration to marketing real-time power demand side management system, trading platform and other related platforms in accordance with relevant rules, and supports the import, saving, declaration and export of declaration data.

7.6.5.2 Aggregate Resource File Management

Virtual Power Plant File Management

Realize virtual power plant file query, file modification, file synchronization and adjustable resource binding maintenance and other operations. Virtual power plant files include virtual power plant name, operation status, virtual power plant type, virtual power plant abbreviation, power generator code, enterprise name, business nature, mailing address, postal code, legal person name, common contact person, office telephone, cell phone number, fax number, e-mail address, enterprise website and other information.

Aggregate Resource File Management

Realize virtual power plant access resource file management and response capability view, including file addition, change, audit and other functions. Response capacity view includes response capacity—peak shaving (kW), response capacity—valley filling (kW), whether to participate in rapid response, rapid response capacity—peak shaving (kW), rapid response capacity—valley filling (kW), rated capacity of the unit (kVA) and other information.

7.6.5.3 Resource Aggregation Simulation Management

Resource Aggregation Grouping

Centralized management of various types of access to virtual power plant resources, providing resource aggregation grouping management, supporting cross-domain aggregation grouping maintenance according to province/municipality/county; aggregation grouping maintenance according to resource characteristics such as interruptible, transferable, transferable, etc.; and aggregation grouping maintenance according to response parameters such as adjustment speed, adjustment duration, climbing time, etc., realizing load resource management at all levels of dimensions.

Load Characterization

Load characterization of virtual power plant access resources, providing load analysis of distributed resources on daily, weekly, monthly, quarterly and annual time scales.

Provide distributed resources with the most energy consumption time, peak and valley load generation time, average load, curve comparison and so on.

Support monitoring of resource operation capability, including resource rated power, feed-in power, output curve, actual power generation, planned power generation; support monitoring of resource load characteristics, including daily maximum, minimum, average load, and peak and valley load monitoring, and on the basis of which, resource load ratio analysis and resource dynamic response capability analysis are carried out again.

Resource Forecast Analysis

Through intelligent processing and pooling of various data, selecting the user's optimal algorithms and parameters through algorithm training, using historical data and intelligent algorithms to forecast loads for users, supporting multi-frequency forecasts such as 96 points within a day. It collects real-time load and predicts customers' future load.

Dispatch Potential Analysis

Based on massive operation data, combine resource type, resource attribute, resource adjustability, resource adjustment potential, load characteristic analysis, and carry out the construction of park source network, load and storage full resource element model.

A multi-dimensional evaluation index system including response speed and response reliability is established, the distribution of scheduling capability on the time scale is considered, an evaluation model considering timing constraints is established, and the minimum set of input parameters of the evaluation model is analyzed by using data dimensionality reduction and correlation analysis for the different operation scenarios of normal/emergency, so as to put forward a multi-timescale online evaluation model for the scheduling capability of different aggregated resources of the virtual power plant.

Evaluation Model

Based on the regulation potential analysis model, the regulation potential of different aggregation units is obtained, including whether they are adjustable or not, the adjustable time period, the maximum output power, the minimum output power, the maximum number of equipments, the minimum number of equipments in different time periods, and so on. Get a clear picture of their response potential and response time periods, etc.

Resource Capacity Measurement

You can manually select resource types, resource attributes, resource adjustable characteristics, load characteristics to analyze load capacity, load response rate, resource ramp speed and other indicators to carry out load resource capacity calculation. Invite customers with controllable loads that meet the criteria for load resource capacity measurement, and incorporate customers who have signed load aggregation contracts into the virtual power plant resource pool management.

Potential Resource Mining

Based on the user load characteristics, energy consumption and energy utilization rate, combined with the results of resource prediction analysis, users are classified and screened to form a potential customer resource pool and realize the management of potential load customers; on this basis, typical cases are directed to be pushed in combination with the potential prediction and the stage load characteristics, energy consumption and other factors.

Resource Aggregation Simulation Program

Transaction program to determine the program resource list prioritization. Determine each resource response capacity allocation. Provide resource optimization recommendations. Classify and aggregate load resources such as building air conditioners, energy storage power stations, etc. Based on the interactive needs of different scenarios, manually select load resources under different characteristics such as response rate, regulation length, regulation capacity, etc., filter and screen the resources through the system, determine the priority order of the resource list, and compare to derive the optimal matching load resources of the current scenario to realize the simulation through the transaction scenario simulation.

Load Resource Monitoring and Evaluation

Based on the dynamic management mechanism of load resources, the system regularly monitors and evaluates the load resources in the pool with indicators such as load capacity, controllable load capacity, efficiency rate, regulation duration, etc., and issues load resource monitoring and evaluation reports, which serve as the basis for judging the release of load resources from the pool and aggregation management.

Market Declaration Model

Provide the quotation mechanism and declaration strategy for participation in the power market, and construct market declaration models for different aggregation units to meet the grid scheduling and market demand and maximize the interests of aggregation units.

Plan Issuance

Import, display and query the plan curve or transaction clearing result issued by the power grid.

Plan Execution

Based on the plan curve or transaction clearing result issued by the power grid, the system provides the plan execution function, including plan curve display, plan curve analysis, plan curve comparison and so on.

7.6.5.4 Collaborative Interactive Response Management

Intelligent Regulation Model

Based on the charging load forecast, system dispatch demand, power market information and other information, and taking into account the response range, rate, time period and other factors, study the intelligent regulation and control strategy of virtual power plant resources participating in grid operation, strengthen the information gathering, data analysis and information flow capability of the load aggregation application, and construct the multi-timescale coordinated control model of pre-date optimization—intra-day rolling—real-time balance to realize the interaction between the virtual power plant and the grid and realize the intelligent decision-making to meet the needs of the power grid. It also realizes intelligent decision-making of virtual power plant interacting with the power grid, so as to meet the demand of power grid dispatching and maximize the interests of users at the same time.

Control Strategy Management

Based on the intelligent regulation model, different decomposition strategies are configured according to virtual power plant resource types, resource attributes, resource adjustability, resource adjustment potential, load characteristics analysis, and different decomposition strategies are flexibly configured according to resource types, regions, aggregation units and other dimensions.

Policy Recommendation

According to the characteristics of the response capability of each scenario of load aggregation, it provides the platform with a strategy allocation plan that takes into account the reliability, economy and fairness of the plan.

Strategy Trial Calculation

Provides response capacity and response speed estimation function for each allocation scheme to provide support for scheme formulation.

Modification of Allocation Rules

Based on the query content, it can modify the pressure drop quota for each scenario from the load gap of different dimensions, such as day-ahead, intra-day and real-time.

Historical Strategy Log

Historical strategy information can be queried by time, scenario name and scenario number, and historical strategy information can be copied and used for current strategy preparation, and historical strategy formulation can also be categorized and counted according to indicator load requirements.

Real-Time Interactive Monitoring

Fully display the results of regulation and control of various types of load resources in the virtual power plant, including information on the plan issued by the power grid, control baseline, real-time power, regulation volume, gap volume, etc., and realize visualization display.

Interactive Response Analysis

Provide statistics on the completion of daily, monthly and yearly multi-timescale regulation of distributed resources; provide analysis on the regulation capacity and level of distributed resources.

7.6.5.5 Effect Evaluation Judgment

Effect Evaluation Model

After the strategy implementation is completed, the implementation effect needs to be evaluated. From multiple dimensions such as actual regulation amount, regulation rate, response time and duration, the load-side resource regulation performance evaluation model is constructed to pave the way for settlement and sharing.

Response Effectiveness Determination

According to the load response standard, the effectiveness of the actual regulation amount, regulation rate, response time and duration of the user demand response is judged; the number of times the user participates in the response and the rate of meeting the standard are displayed in each scenario, and the detailed information on the response situation at the user level can be displayed and exported.

Operator Effectiveness Evaluation

Quantitatively evaluate the operators in multiple dimensions, such as theoretical regulation power, actual regulation power, regulation power percentage, regulation rate, response time, peak time and duration, and then give the evaluation grade to actively guide the operators to carry out optimization and transformation.

Black and White List Management

For aggregation resources with low response efficiency, blacklist management is carried out to eliminate such sites at the time of ordering.

7.6.5.6 Revenue Settlement Management

Settlement Data Preparation

Provides settlement data synchronization function. The settlement data preparation function realizes the verification of the required data before each settlement to ensure the confirmation of the preliminary data before settlement and the accuracy of the subsequent settlement. The system provides daily, monthly and yearly time scale data display of power consumption and regulation of each distributed resource.

Revenue Measurement

Based on the data confirmed in the settlement data preparation process, the system performs revenue calculation on the specified transaction results in accordance with the trading rules, and displays the results of revenue calculation, final settlement results, and deviations.

Expenses

Provides details of customer expenses such as energy consumption and operation and maintenance fees. Provide multi-dimensional statistics on the trend of expense expenditure and the proportion of expense expenditure.

Resource Sharing and Settlement

Provides management of monthly settlement data issued by the ancillary service market in each month. Provide monthly calculation of each resource user's revenue; provide user's reconciliation according to different data sources; revenue sharing, and calculate the sharing ratio and revenue according to the revenue sharing rules.

Comprehensive Customer Settlement

Provide new customer monthly settlement function according to the synthesis method. Completes the monthly mileage adjustment and monthly revenue calculation of each resource user on a monthly basis. Provides the function of issuing settlement statement; provides the function of user monthly settlement information. Provide the function of archiving the monthly settlement of customers.

Market Transaction Review

It is mainly used to synchronize the operation of the virtual power plant resource pool in real time, track the clearing results and response effect of the virtual power plant in the process of participating in market transactions, and analyze the response effect, strategy formulation and implementation, and single response gain and loss from the market, strategy, and gain/loss perspectives.

7.7 Pilot Sample Design

7.7.1 Design Logic and Workflow

In this section, demonstrate the process of completing the design and the resulting outcome through a pilot example design.
Problem Definition (P)

Core Problem
How to integrate distributed energy resources(DERs) such as solar PV, battery storage, electric vehicles (EVs), and flexible loads into a coordinated "virtual" power plant that enhances grid stability, reduces energy costs, and participates in electricity markets?

Objectives

- Aggregate DERs into a controllable virtual generation unit.
- Dynamically respond to grid demands (e.g., peak shaving, frequency regulation).
- Maximize revenue through market participation.

Input Design (I)

Input Components

1. **Resource Data**
   - Solar PV power forecasts (weather-dependent).
   - Battery state of charge (SOC), capacity, and charge/discharge limits.
   - EV charging demand and flexible scheduling windows.
   - Industrial/commercial load flexibility curves.
2. **External Signals**
   - Grid operator commands (e.g., frequency regulation signals).
   - Electricity market prices (real-time pricing, peak/off-peak rates).
   - Weather forecasts (for solar generation prediction).

Logic Design (L)

Core Algorithms

1. **Resource Aggregation**
   Model DERs as "dispatchable power nodes" and compute total available capacity and flexibility ranges.
2. **Optimization**
   Objective Function

   $$\text{Minimize Cost (or Maximize Revenue)} = \text{Market Income} - \text{Penalties (e.g., constraint violations)}$$

   Constraints
   - Battery SOC limits: $SOC_{min} \leq SOC \leq SOC_{max}$.
   - Solar generation bounded by weather forecasts.
   - Loads must meet user requirements.
3. **Market Strategy**
   - Discharge batteries during high-price periods, charge during low-price periods.
   - Bid capacity into ancillary services (e.g., frequency regulation).

Output Design (O)

System Outputs

- **Real-time Dispatch Signals**
  Example: *{"battery_storage": "charge_at_100kW_12:00-14:00"}*.
- **Market Transactions**
  Volume and price of energy traded in markets.
- **Grid Impact**
  Peak load reduction (e.g., 10% lower demand), carbon emissions savings.

Visualization (Dashboard)
VPP dashboard showing real-time DER status, revenue streams, and grid performance.

7.7.2 General Design Approach and Design Tools

First confirm the external conditions and basic indicators identified in the planning phase as the original basis for the design.

Confirming and selecting the configuration of the distributed energy and smart metering control units.

Then designing and selecting the equipment and parameters of the communication layer.

Then design the data layer on top of the communication layer to realize the design of system architecture, data collection and data governance methods, network and network security, system integration, etc., following the project specifications and requirements and confirming the system security level.

Finally, an interactive interface is used to realize the display of aggregated resources, including various distributed energy resources aggregated by the virtual power plant, such as photovoltaic power generation, wind power generation, and energy storage systems. This function displays the geographic location, installed capacity, real-time output and other information of various types of resources in the form of charts and maps, etc., to help users understand the distribution and operation status of the resources.

In the aggregated resources interface need to include.

Aggregation optimization, showing the situation of each virtual unit and the optimized combination;

Potential Assessment, which assesses the user's capacity, as well as the regulation potential of the unit and the plant as a whole

Demand Response, to view demand response participation.

Finally, power trading and provision of ancillary services of the virtual power plant need to be realized according to the rule constraints and technical conditions of the power trading market.

7.7.3 Proposed Technologies

In the sample design, centralized architecture is used with cloud-accelerated intelligent energy management and control. Here is proposed key technologies in this configuration.

7.7.3.1 IOT Collection for VPPs

Based on the "cloud, pipe, side and end" IOT collection architecture, the intelligent energy management and control and virtual power plant platform is constructed. By installing metering terminals, collection terminals and other equipment on the end side, installing intelligent energy gateways and other equipment on the side side, generating optimized control strategies on the cloud side, and sending them to the side and end for execution through the pipe layer, it realizes the coordinated and optimized control of multiple energy sources, improves the efficiency of energy use, and reduces the enterprise's energy use cost.

The cloud connects with the side and end through the pipe, the side connects with various types of ends of the same type, the end is divided into collection and control, and at the same time provides the basis and execution of control for the side and the cloud, the cloud unifies the regulation and control, collects massive energy information in normal times, and the side controls the energy supply and use equipments through the optimal dispatch to improve the energy efficiency.

Cloud

Massive data are aggregated to the cloud, with data processing, integrated analysis and effect assessment and other functions, such as data mining, machine learning, and big data analysis, to support intelligent energy management and control. The cloud is also responsible for optimization strategy generation and supply and use coordination and balance management.

Pipe

Including Internet, optical fiber, GPRS virtual private network, 4G private network, 5G private network and other communication pipes, which can support both wired and wireless communication.

Edge

Local edge router deployed in the workshop, power station house and other substations, including the implementation of local energy optimization and control strategies, through edge computing can achieve real-time, short-cycle data analysis, to better support the timely processing and implementation of the local business, is to achieve multi-energy complementary, energy efficiency is an important basis for the improvement.

End

Containing collection and control terminals, the collection contains measuring instruments and various types of sensors.

7.7.3.2 Communication Architecture

Physical architecture

According to the requirements of the controllable load of the virtual power plant, the internal network is divided into production control area, management information area and Internet in accordance with the relevant network security requirements. The specific deployment logic is as follows:

Physical Architecture Diagram

1. Network communication with the Virtual Power Plant Operation and Management Center is divided into two categories, one involving control-type commands directly initiated by the Virtual Power Plant Operation and Management Center, which needs to be accessed through a special vertical encryption device, via the operation leased line mode, and through a special edge gateway to the platform; and the other is monitoring and management-type data, which is accessed via the Internet.
2. Network communication with the power trading center is accessed to the platform through the Internet in accordance with relevant requirements.
3. With user-side distributed photovoltaic, decentralized wind power, distributed energy storage and user load system, accessed by system-to-system or smart energy gateway through the Internet.

Hardware Design, Intelligent Gateway

Intelligent gateway is used in central air conditioning, distributed photovoltaic, decentralized wind power, distributed energy storage and other distributed power supply, load regulation and control devices to provide remote telecommunication, telemetry, telecontrol, telecontrol, and automatic control to ensure stable operation of the power system with the goal of monitoring the stable operation of the power system in the steady state and transient state.

1. Technical realization Technical realization: based on embedded architecture, it is mainly applied to the conversion of industrial/enterprise field equipment protocols, realizing data acquisition and forwarding of different protocols, meeting

the relevant standards and norms of electric power safety, and being a communication bridge between the plant station and the main station, realizing the data from the plant station to all levels of the dispatching system or the group's unified monitoring and operation and maintenance platform.

2. Interface Provides a wealth of interfaces, including: 2 6 Ethernet ports, 6 8 RS485/RS232 serial ports, built-in wireless 4G/5G communication interfaces, AC and DC power supply interfaces, to meet the needs of a variety of application scenarios.
3. Processor industrial-grade microprocessor as the core, high processing speed, able to collect accurate measurement data.
4. Protocol RS485 bus and Ethernet channel, support MODBUS protocol, DL/T 645-2007 Multifunctional Energy Meter Communication Statute, 104 protocol, and private protocols of equipment manufacturers, and it can collect data from many kinds of meters. It is also equipped with DI and DO switching control interfaces, which can remotely monitor the valve control of AC and DC loads.
5. Wireless communication The wireless channel can communicate with the remote data center, upload monitoring data and control equipment status through wireless network. The wireless channel can also use 4G data transmission module and APN special network card to communicate with the special network data center and upload monitoring data, equipment status and other information.
6. Data storage external multiple storage methods to save massive data. Moreover, the device has hardware encryption interface and supports various encryption methods to provide security for data transmission.

Pipe layer, using stable, reliable and safe communication methods for communication and data transmission, to ensure information security.

Side layer is to realize the functions of data collection and data transmission. End layer for distributed generation resources, energy storage, and terminal equipment for controllable loads.

Cloud layer, responsible for data processing, analysis, and control, and provides intelligent energy management and control services to the end user.

7.7.3.3 "Big Platform, Micro Service" Cloud Platform Architecture

Based on "big data, cloud computing, micro-application and mobile Internet", it is constructed and designed with resource elasticity expansion capability; application horizontal expansion capability; distributed data storage capability; structured and unstructured data storage, searching and mining capability; big data integration and computation capability; data display visualization capability; data security service capability; data deletion capability for key data; and data deletion capability for key data. Capability; data deletion and modification record function for key data, and the ability to record the operation history of key data at any time. It provides a basic development framework, including basic development components: mobile components, internationalization, data access, transaction processing, exception control, access control, etc., which facilitates the development of business applications that can be easily uploaded to the cloud; it provides a variety of basic services for developers to use, and technically adopts the internationally mature servitization framework, which provides a full set of basic services, including: application containers, load balancing, databases, Caching, storage, logging and other services, in the micro-application business development, developers can pay more attention to the realization of business functions, such as 4A, security, performance, data storage, distributed implementation of the development of difficult, key points of the technical processing, the platform to provide a unified solution, so that the business micro-application developers can pay more attention to the realization of the business functions to enhance the development efficiency, improve the security and reliability of the cloud platform. Security and reliability (Fig. 7.3).

Cloud Platform Architecture

Based on advanced open source technology, adopting the latest design ideas of cloud architecture and micro-application, divided into IAAS, PAAS and SAAS three-tier architecture, providing a PaaS platform with cloud application (SaaS) support from software development, validation, to release, O&M and operation of the complete lifecycle, it can help enterprises establish a unified micro-application architecture and ecosystem, and provide them with an open cloud platform solution for enterprises.

IAAS Layer

The cloud platform can be built with different IaaS platforms, supporting the IaaS platforms of infrastructure and operation service providers, as well as virtualization technology platforms, while it can be deployed directly on physical machines. It ensures isolation from hardware and supports different hardware models.

The Yuxi Cigarette Energy Management System will be installed and deployed on the IAAS layer resources provided by Redtag Group, and relevant security technologies are used to guarantee the security isolation from the virtualized core data resources, the database uses the open source MYSQL software, the structured storage uses the mongoDB software, the application cache uses redis, and the load balancing and reverse proxy installs the Ngnix software.

PASS layer

Cloud development platform provides basic servitization framework, realizes lightweight, modular, servitization (micro-application) technology system, provides basic development framework, distributed caching service, distributed database access service, distributed message queuing service, distributed text retrieval service, third-party payment, data collection service, 4A security service, resource monitoring service, workflow service, Task scheduling service, messaging service,

Fig. 7.3 Summary of the big platform, micro service structure

cloud mail service, cloud report service and other functions, see the description in the technical platform implementation.

SAAS layer

According to the requirements of this project, the basic resources and services provided by the IAAS and PAAS layers are utilized, and the enterprise application development platform is applied to simplify the development of intelligent energy management business applications.

Front-End and Back-End Separation Technology

Front-end and back-end separation has become the industry standard for Internet project development, and the decoupling of front-end and back-end is effectively carried out through nginx+tomcat. At the same time, front-end and back-end separation also lays a solid foundation for the project's distributed architecture, elastic computing architecture, microservice architecture, and multi-end services (multiple clients, such as browsers, in-vehicle terminals, Android, IOS, etc.).

The front-end and back-end of the system are separated using the VUE framework realized by MVVM design thinking as the front-end basic framework, and realized based on SPA (Single Page Application) single page development mode. The architecture of front-end and back-end separation is shown in the figure below (Fig. 7.4):

(1) Interaction form

    The back-end only needs to agree with the front-end to receive and return data format (usually in JSON format), to provide API interfaces to the front-end. The front-end can call the API through HTTP requests for interaction. After the front-end to obtain the data, page assembly, rendering, and ultimately rendered in the browser.

(2) Code organization

    Front-end and back-end code is placed under different projects, the front-end code can be developed independently, through mock technology to simulate the back-end API services can be run independently, testing; back-end code can also be developed independently, running, testing, through swagger technology can automatically generate API documents for the front-end to read, but also automated interface testing, to ensure that the availability of the API to reduce the risk of integration.

Big Data Platform

The overall system big data platform is built based on cloud technology component capabilities, and the platform architecture is designed from the multi-layer structure

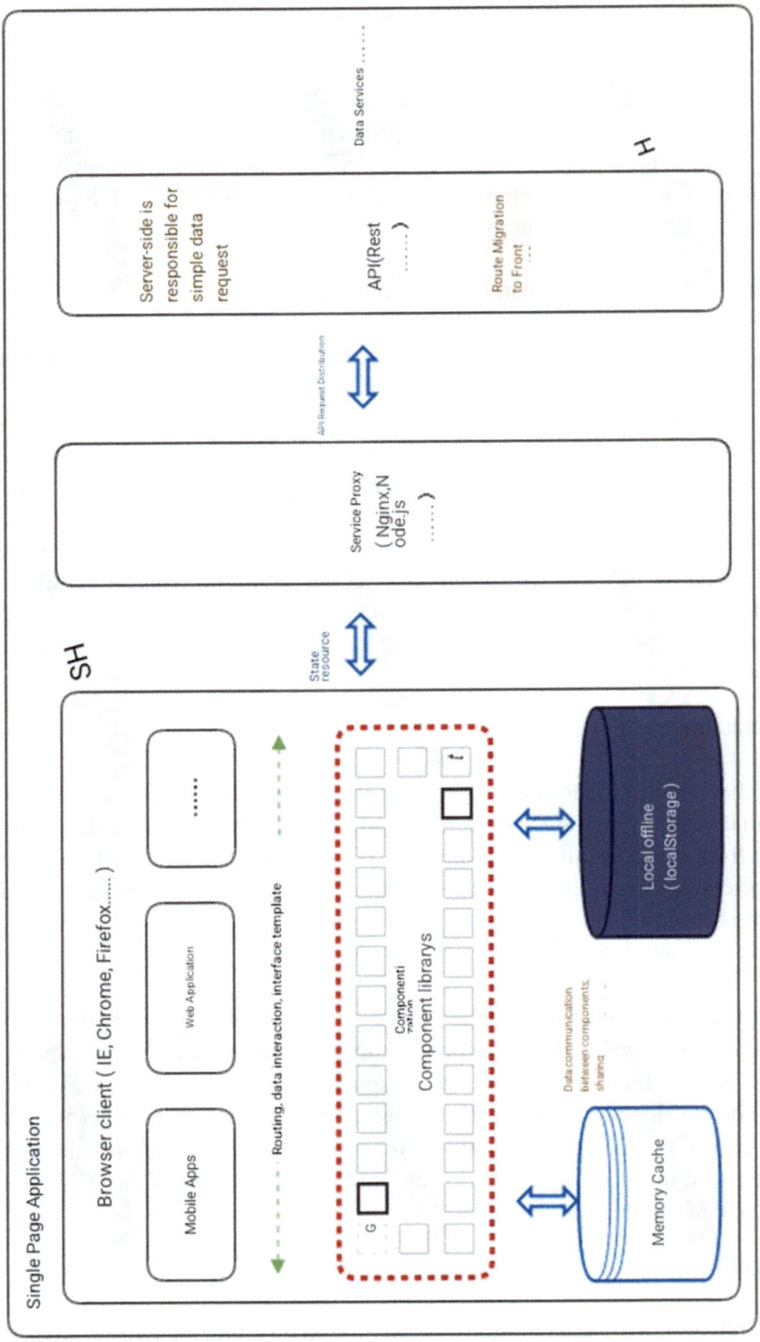

Fig. 7.4 Front-end and back-end of the system structure

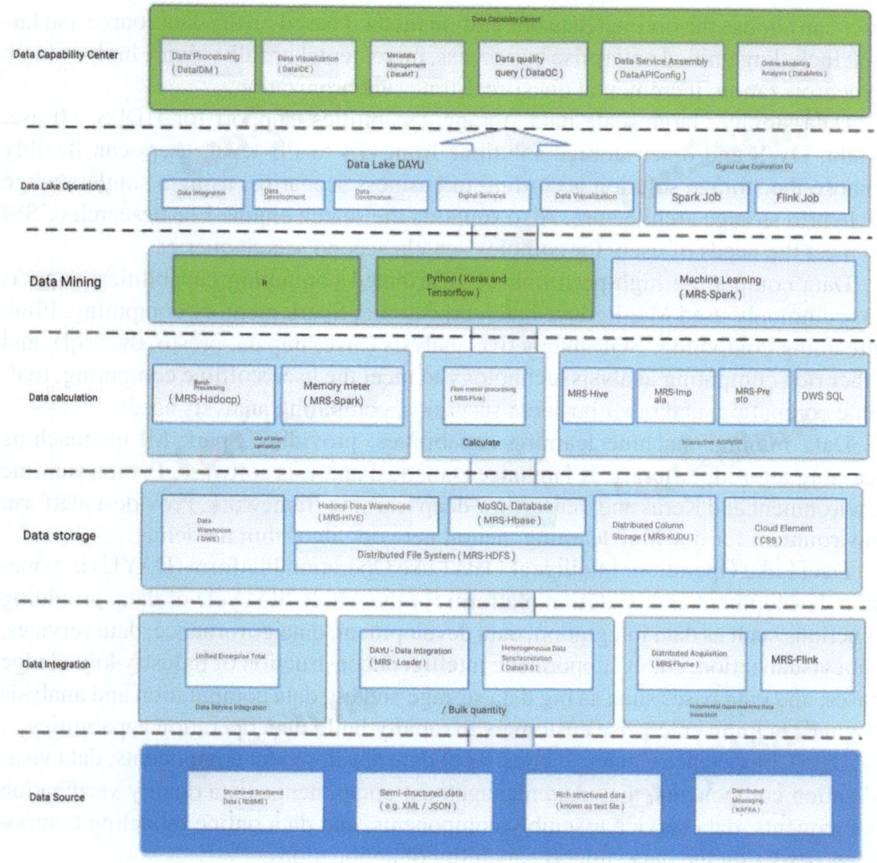

Fig. 7.5 Big data platform

of data source, data integration, data storage, data computing, data mining, data lake operation, and data competence center. It establishes unified data integration, large-scale data storage, and integrates the platform capabilities of big data processing, computing and data services to effectively reduce the cost of big data usage. As Fig. 7.1.

Data Processing Flow

Data source: rich data source types. Data sources support commonly used RDBMS relational libraries, such as Oracle, MySQL, SqlServer, GBase, etc.; support for semi-structured data, structured data; support for Kafka distributed messaging system quasi-real-time subscriptions (Fig. 7.5).

Data Integration: Multi-source high-performance data integration capabilities to meet a variety of heterogeneous data, high-performance data synchronization, the user can choose the optimal data integration method based on the data source and target; including unified enterprise bus access, incremental or full-volume high-volume synchronization, incremental quasi-real-time synchronization.

Data storage: large-scale data storage capabilities, support for HDFS, HBase, Kudu, DWS and other storage solutions from GB to PB level, users can flexibly choose the storage solution according to business scenarios, using a single storage or hybrid storage architecture. Also supports the search engine Elasticsearch (CSS) to meet the needs of users for complex search, second search queries.

Data computing: high-performance distributed computing capabilities, support for commonly used MapReduce batch processing, Spark memory computing, Flink streaming computing, SQL interactive analysis (hive, impala, presto, dws sql), and other rich computing analysis technology to meet the user's offline computing, real-time computing, and other business scenarios, computing analysis needs.

Data mining: machine learning capabilities, providing Spark MLlib machine learning algorithm library, R language statistical analysis software, Python runtime environment and Keras and Tensorflow deep learning framework. Provide a platform environment for machine learning, neural network algorithm modeling.

Data Lake Operation: Intelligent Data Lake Operation Platform (DAYU) is a one-stop development and operation platform for the whole life cycle of data, providing functions such as data integration, data development, data governance, data services, data visualization, etc. It supports the intelligent construction of industry knowledge bases, and data bases such as big data storage and big data computation and analysis engines, helping enterprise customers to rapidly build data operation capabilities.

Data Competence Center: It consists of data development components, data visualization components, metadata management components, data quality verification components, data service assembly components, and data online modeling components, realizing the data support capability of applications.

Data Processing Component

Data Development Component: Chassis data processing DataIDM realizes front-end one-stop, closed-loop and formulated big data development and management services including large-scale data integration, data calculation, workflow scheduling, task scheduling, etc.; it is agile and easy to use, mature and stable, safe and reliable, open and flexible, and provides services such as high-efficiency data synchronization, high-performance distributed computation, and search engine, etc.

Data Visualization Component: Data Visualization DataIDE provides as simple and easy-to-use visualization interface, based on simple mouse point and click, drag and drop, etc., for the data-related personnel without technical background to self-service rapid data query, analysis and data drilling, and even for the rather scattered and complex data sets to carry out in-depth mining and analysis on business.

Metadata Management Component: DataMT, a metadata management service, is dedicated to processing technical metadata, business metadata, and management metadata to help users gain more insights into the data, and then tap into the value hidden in the data.

Data Quality Verification Component: DataQC, a data quality verification service, is based on the target data set or source database, and the user verifies the quality of the target data by formulating different dimensions of the data quality verification method.

Data Service Assembly Component: Data Service Rapid Assembly Service DataAPIConfig is a low-code development platform that can quickly configure the development of enterprise-level API interfaces, which does not require a basic knowledge of programming can be quickly started, and in a short period of time, the data can be realized as a microservice-based API interfaces.

Data online modeling components: online modeling and analysis services DataMetis set of data processing, machine learning algorithms call, model training, model effect evaluation, model deployment as one of the one-stop modeling and analysis platform. Relying on distributed storage and computing power, it has built-in rich machine learning algorithms and formed a set of mature modeling and analysis training system, which provides modeling and analysis users with easier operation experience and stronger processing performance.

7.7.3.4 Smart Function Module

Distributed Energy Synergy

Building roofs and surrounding open space are suitable for the development of distributed PV. The virtual power plant platform unifies the control of PV power stations in the plant and dynamically adjusts the PV output according to solar irradiation and real-time load. If there is sufficient light and electricity consumption in the plant, the PV power will be stored in the supporting energy storage system as a priority; if the electricity consumption is tight, the PV power will be directly supplied to the production, and then the power will be taken from the grid when there is insufficient power, thus enhancing the proportion of self-generated power and self-consumption, and increasing the amount of clean energy consumed by the plant by the level of megawatt hours per year, and reducing the amount of carbon emission is also especially considerable, so as to realize the transformation of green production.

Specific functions include:

1. Resource inventory and integration
   Comprehensive inventory of resources, accurate mapping of plant roof area, orientation and carrying capacity, planning for distributed photovoltaic arrays; sorting out the potential for waste heat and pressure recovery in aluminum processing, assessing the suitability of waste heat boilers, turbines and other equipment; statistics on the layout of unused vacant land for small-scale wind power or biomass energy facilities, and unified access to dispersed energy resources in the virtual power plant control system, to build a resource pool that covers all types of distributed power generation, storage and waste heat utilization devices, realizing

a one-stop resource pool for the utilization of energy. The resource pool covering various types of distributed power generation, energy storage and waste heat utilization devices is constructed to realize one-stop management.

2. Intelligent power generation scheduling
Based on the weather forecast, historical power generation data and real-time electricity load, the virtual power plant platform dynamically generates distributed energy generation plans. When there is sufficient light, priority is given to increasing photovoltaic output, and the excess self-consumption portion is stored or sold to the power grid; when the wind is suitable, the wind turbine speed and pitch angle are optimized to stabilize wind power; and the waste heat power generation is flexibly deployed in accordance with the heat balance of the production process to ensure that the energy output is accurately matched with the complex and variable power consumption curves, to increase the self-generation and self-consumption rate to more than 80%, and to reduce the dependence on purchased power.
3. Complementary energy storage linkage
Supporting lithium batteries, super capacitors and compressed air energy storage and other multi-dimensional energy storage systems, the virtual power plant gives it the key mission of "shaving peaks and filling valleys, smoothing out fluctuations". When the load is low, the energy storage absorbs the surplus electricity; during peak hours or grid fluctuations, the energy storage instantly discharges to ensure uninterrupted power in key production processes and maintain voltage stability. Combined with the fluctuating characteristics of distributed energy generation, energy storage intelligently switches the charging and discharging modes, such as storage when the PV is large and the load is not yet full, and release when the PV decreases abruptly and the load increases suddenly, so that the standard deviation of the power supply of the plant area can be reduced by 30–50%, and the resilience of the energy supply can be strengthened.
4. Energy mutual aid and emergency protection
Build internal energy LAN, virtual power plant to coordinate energy supply and demand, one energy surplus to make up for the shortfall in other places; in case of grid failure, quickly switch to the "island mode", distributed energy and energy storage to support the core production, to ensure continuous operation without stopping the production, to maintain the basic production capacity of 60–80%, to be seamlessly restored to the grid and then connected to the grid, to enhance the energy autonomy and disaster resilience of enterprises. This improves the enterprise's energy autonomy and disaster resistance, and realizes a green, efficient, safe and reliable production pattern.

Flexible Production Scheduling

After the virtual power plant platform is accessed, external factors such as electricity price signals and grid frequency adjustment demand are incorporated into the production scheduling logic. When the grid frequency fluctuates, the platform quickly calculates and instructs part of it to briefly adjust the working condition, participate in a frequency adjustment, and respond within a few seconds to provide megawatt-level reactive power support, safeguard the stability of the power grid, and the enterprise is compensated for ancillary services, expanding revenue channels, and deeply integrating itself into the ecology of interaction between supply and demand of electric power.

1. Interruptible load sorting in production process
   Deeply analyze the production process, precisely locate the links and equipment with flexible adjustment potential, and list some non-critical assembly lines and auxiliary material handling equipment that can be shut down for a short period of time as interruptible load resources. Quantify the duration of interruption, recovery time, power adjustment range and other parameters of each equipment to build a detailed list of interruptible loads, which is entered into the virtual power plant scheduling system, laying the foundation for flexible regulation and control.
2. Multi-source data fusion drives scheduling decisions
   Integrate multiple data such as production plan, order delivery progress, equipment health status, energy market price fluctuations, real-time grid operating conditions (frequency, voltage, load margin). With the help of big data analysis and intelligent algorithms, it generates flexible production scheduling plans in real time based on different priority weights (e.g., to protect the production of key orders, to pursue the optimal cost of energy, and to assist in emergency regulation of the power grid), and realizes the dynamic adaptation of the production rhythm to the supply of energy and the demand of the power grid.
3. Real-time response to grid regulation commands
   When the grid frequency deviation, voltage drop or emergency overload, the virtual power plant system instantly analyzes the instructions, accurately locates the interruptible load resources in the aluminum industry, and issues regulation and control instructions at millisecond level. For example, it instantly cuts specific power by 10–30%, suspends non-emergency casting processes, synchronizes and coordinates distributed energy generation increment and energy storage discharge, quickly injects reactive power or active support into the grid, ensures grid stability, gains ancillary service revenue based on response contribution, and deeply integrates into the interactive ecology of power supply and demand.
4. Optimized management of production recovery
   After the grid disturbance is lifted, the virtual power plant arranges the production recovery sequence in an orderly manner according to the length of equipment downtime, the remaining order delivery timeframe, and energy cost considerations. The machine learning prediction model is used to estimate the energy consumption and capacity creep curve of the recovery process, and reasonably allocate energy, such as prioritizing the restarting of key equipment at full power and the gradual resumption of auxiliary processes, so as to achieve a smooth transition from emergency regulation to normal production, keep the loss of production interruption to a minimum, and maintain the operational efficiency and economic benefits of the aluminum industry.

Electricity Price Forecasting

Electricity price prediction is related to the power generation plan as well as revenue, and the accuracy of electricity price prediction is realized through the following functions:

1. Data collection and pre-processing

   (1) Extensive data collection: Collect historical electricity price data in the electricity market, including electricity price information in different time scales (e.g., hourly, daily, monthly, and yearly), as well as various types of data related to electricity price, such as electricity load, generation capacity, fuel prices, weather data, and changes in policies and regulations.
   (2) Data cleaning and standardization: the collected data are cleaned to remove outliers, missing values, etc., and then standardized to make data from different sources and of different magnitudes comparable and to provide an accurate and consistent data base for subsequent analysis and prediction.

2. Feature Extraction and Selection

   (1) Feature engineering: extract features related to electricity price from a large amount of data, such as cyclical features (daily cycle, weekly cycle, seasonal cycle, etc.), trend features, fluctuation features, and features closely related to supply and demand in the electricity market, such as load peak and valley differences, availability of power generation equipment, and so on.
   (2) Feature selection: Using statistical methods and machine learning algorithms, the extracted features are screened to select the features that have a significant impact on the electricity price prediction, reduce the data dimensionality, and improve the efficiency and accuracy of the prediction model.

3. Prediction model construction and training

   (1) Model selection: According to the characteristics of the electricity price data and prediction needs, select the appropriate prediction model, such as time series model (ARIMA, VAR, etc.), machine learning model (linear regression, decision tree, random forest, neural network, etc.) or deep learning model (LSTM, GRU, etc.).
   (2) Model training: the preprocessed data is divided into training set and test set, and the selected prediction model is trained using the training set, so that the model can fit the historical tariff data as accurately as possible by adjusting the parameters and hyperparameters of the model, and learn the laws and patterns of tariff changes.

4. Electricity price prediction and result evaluation

   (1) Real-time electricity price prediction: Input real-time data into the trained prediction model to predict the future electricity price, and output the prediction results, including the predicted value of the electricity price and the confidence interval of the electricity price, etc., to provide a reference basis for the operation decision-making of the virtual power plant.
   (2) Evaluation of prediction results: Adopt appropriate evaluation indexes, such as Root Mean Square Error (RMSE), Mean Absolute Error (MAE), Mean Absolute Percentage Error (MAPE), etc., to evaluate the prediction results, analyze the accuracy and reliability of the prediction model, and optimize and improve the model based on the evaluation results.

5. Scenario analysis and risk warning

(1) Scenario setting: Considering different market scenarios and risk factors, such as power market reform, policy adjustments, extreme weather events, energy price fluctuations, etc., a variety of scenarios are set up to forecast and analyze electricity prices under each scenario.
(2) Risk Alert: Based on the results of scenario analysis, identify the risk points and uncertainties of electricity price fluctuations, and send risk alert messages to virtual power plant operators in a timely manner to help them formulate corresponding risk response strategies and reduce operational risks.

Load Forecasting

The virtual power plant load forecast function is crucial to the stable operation and optimal dispatching of the virtual power plant, and the following are its main functional features:

1. Data Collection and Integration

   (1) Multi-source data collection: Comprehensively collect load-related data, including historical load data, meteorological data (e.g., temperature, humidity, wind speed, etc.), date types (weekdays, holidays, etc.), economic data (e.g., regional gross domestic product, industrial value-added, etc.), as well as real-time grid operation data, etc., in order to provide a rich data base for load forecasting.
   (2) Data cleaning and preprocessing: The collected data of various types are cleaned to remove noise, outliers and missing values, and preprocessing operations such as standardization and normalization are carried out to ensure the quality and consistency of the data and improve the accuracy of the forecast.

2. Feature engineering and selection

   (1) Feature extraction: Extract features related to load changes from the original data, such as daily, weekly and seasonal characteristics of the load, as well as characteristics of the impact of meteorological factors on the load, etc., to explore the potential laws and correlations in the data.

(2) Feature screening: Using statistical analysis, correlation analysis and other methods, we screen out the features that have a significant impact on load forecasting, remove redundant information, reduce the data dimension, reduce the amount of computation, and at the same time improve the performance and efficiency of the forecasting model.

3. Prediction model construction and training

(1) Model selection: According to the characteristics of the virtual power plant load data and forecasting needs, choose a suitable forecasting model, such as time series model (ARIMA, VAR, etc.), machine learning model (linear regression, decision tree, random forest, support vector machine, etc.) or deep learning model (LSTM, GRU, etc.).
(2) Model training and optimization: divide the preprocessed data into training set and test set, train the selected prediction model using the training set, make the model fit the historical load data accurately by adjusting the model's parameters and hyper-parameters, and validate and optimize it on the test set to ensure the model's generalization ability and prediction accuracy.

4. Load forecasting and real-time monitoring

(1) Short-, medium- and long-term forecasting: According to the demand of different time scales, short-term (hourly, daily), medium-term (weekly, monthly) and long-term (quarterly, annual) load forecasting is performed to provide decision support for the virtual power plant's scheduling plan, unit start-up and shutdown arrangements, and maintenance plan.
(2) Real-time Load Tracking and Adjustment: Combining with real-time grid operation data, real-time tracking and forecasting of the current load, timely detection of sudden changes and abnormal fluctuations in the load, and dynamic adjustment of the forecast results, to provide accurate load information for the real-time scheduling and control of the virtual power plant.

5. Evaluation and visualization of prediction results

(1) Performance evaluation: quantitative evaluation of load forecasting results using evaluation indexes such as root mean square error, average absolute error, average absolute percentage error, etc. to analyze the accuracy and reliability of the forecasting model, so as to discover problems and improve the model in time.
(2) Visual display: The load forecast results are visualized in the form of charts (such as line graphs, bar charts, heat maps, etc.), which visually presents the trend of load changes, forecast intervals, and uncertainties, facilitating understanding and analysis by operation personnel and assisting in decision-making.

Decision-Making Analysis The decision analysis function of the virtual power plant is the key support for its efficient, intelligent and economic operation, covering multiple dimensions and aiming to cope with the complex and changing energy market and the dynamics of supply and demand.

1. Decision-making for optimal allocation of energy resources

   (1) Scheduling of power generation resources: Comprehensive distributed energy (photovoltaic, wind power, small hydropower, waste heat power generation, etc.) power generation capacity, cost curve, combined with real-time and forecast load demand, the use of linear planning, dynamic planning algorithms, accurate decision-making of the power generation unit start-stop, output setting, to ensure that the energy output maximizes the matching of the electricity load, to enhance the proportion of self-generation and self-consumption, to reduce the rate of energy abandonment, and reduce the cost of power generation by 15–30%.
   (2) Energy storage charging and discharging strategy: based on time-sharing tariffs, load fluctuations, grid ancillary service demand, through the model simulation of different charging and discharging time combinations of revenue, to determine the optimal charging and discharging time and power of the energy storage system, such as peak and valley arbitrage to ensure that daily charging and discharging cyclic efficiency of more than 80

2. Market trading strategy development

   (1) Electricity and energy trading choices: in-depth analysis of electricity spot, medium and long-term contract market price trends, integration of their own power generation costs, available power estimates, with the help of machine learning prediction model (neural network, random forest), decision-making to participate in the market timing and power quotes, capture high-price window to sell electricity, accurate pricing to enhance the winning rate of the bid, transaction profitability increased by 20–40%.
   (2) ancillary service participation trade-off: assess the ability and cost of virtual power plants to provide FM, VR and standby capacity, combine with the demand for ancillary services released by the grid and the compensation price, quickly weigh participation and the amount of resources invested in balancing the loss of equipment, opportunity cost and revenue, expanding multiple channels of revenue, and enhancing the depth of grid interaction.

3. Risk assessment and response decision-making

   (1) Price fluctuation wind control: real-time monitoring of energy market price sensitivity factors (fuel prices, policies, climate anomalies), Monte Carlo simulation and other means to quantify the fluctuation interval and probability of electricity prices and fuel prices, pre-set contract locks and hedging ratios in advance, to guard against price risks, protect revenue stability, and fluctuations in the loss of control within 10%.

(2) Operational Reliability Guarantee: Based on equipment failure history, real-time health monitoring data, fault tree, Bayesian network analysis of potential failure scenarios and the scope of the impact, the development of preventive maintenance, emergency scheduling plans, such as key power generation equipment failures in advance to shift the load, deploy redundant resources to maintain power supply reliability of more than 95%.

4. Benefit and performance evaluation decision-making

(1) Cost-benefit accounting: We calculate the full costs of investment, operation and maintenance, fuel, and manpower, benchmark power generation, ancillary services, and energy saving and fee reduction, and visually present the profit contribution rate of each business segment, so that we can optimize the direction of resource investment and business expansion to ensure that the return on investment is leading in the industry.

Performance indicator tracking: Setting multi-dimensional indicators such as power generation efficiency, load response timeliness, energy utilization efficiency, etc., and reviewing them on a regular basis to identify gaps in the industry benchmarks, so as to provide precise targets for operation optimization, technology upgrading, and personnel training, and to drive the continuous advancement of the virtual power plant.

Trading Strategy The trading strategy function of the virtual power plant is a key means to maximize economic benefits, optimize resource allocation and stabilize grid operation in the complex power market environment, and has various key elements:

1. Market Price Insight and Forecasting: Integrating massive historical electricity price data, real-time power supply and demand information, energy policy dynamics, and meteorological impact factors on energy output, etc., the company uses time series analysis and machine learning models (such as LSTM neural network) to accurately predict the price trend of electricity and ancillary services. For example, it predicts the fluctuation of the price difference between peak and valley electricity one week in advance, and the error rate is controlled within 5%, which is the foundation for the choice of transaction timing, and helps the virtual power plant to capture the window for selling electricity at high price and purchasing electricity at low cost.
2. Electricity trading optimization: Based on the cost of power generation (including distributed energy operation and maintenance, fuel costs, etc.), the estimated amount of available electricity and price forecasts, we build linear programming or game theory models to customize trading strategies. Weighing the proportion of medium- and long-term contracted power to protect the basic income, and allocating flexible power in the spot market to catch up with the high price difference; according to the characteristics of different power sources, spot selling during the daytime when the PV is large, and selling at nighttime when the power is stored in the low valley, which will improve the profit of kWh power by 20–30%.

3. Precise participation in ancillary services: real-time monitoring of grid frequency and voltage stability indicators, assessment of its own frequency regulation, voltage regulation, standby capacity response speed, regulation accuracy and cost, according to the market demand for ancillary services and compensation rules, the dynamic planning decision to participate in the project and offer. For example, at the moment of rapid load rise in the power grid, it can quickly put in energy storage and uninterruptible load resources, respond to frequency regulation in milliseconds, reasonably offer lucrative compensation, and expand diversified revenue streams.
4. Cross-market arbitrage synergy: Keeping an eye on related markets such as electric energy, carbon trading and renewable energy green certificates to tap arbitrage space. If the virtual power plant has abundant clean energy generation during a certain period of time, in addition to the sale of electricity, it will simultaneously quantify the value of excess green certificates and carbon emission reduction and package them for sale, which will increase the comprehensive income by 15–20%, realizing the linkage of cross-market resources and value.
5. Risk hedging and contract management: In view of the risk of energy price fluctuations and policy changes, the introduction of financial derivatives (e.g., power futures, options) hedging, depending on the risk exposure to set the contract size; intelligent management of trading contracts, from generation, execution to settlement of the entire tracking, early warning of performance risks, to ensure smooth compliance with the transaction, locking in a stable income, fluctuations in the loss of control within the expected range.
6. Real-time trading adjustment: real-time optimization strategy with market conditions, power generation equipment status, and sudden load changes. In case of power generation failure and sudden reduction of output, the company will quickly change the sale of power to purchase power to ensure internal supply and demand; in case of over-expected load, the company will increase the supply of ancillary services, and will flexibly adapt to ensure the balance of operational resilience and benefits.

Transaction Settlement The transaction and settlement function of the virtual power plant is the core module that ensures smooth operation of economic activities, reasonable distribution of interests of all parties, and clear and accurate financial relationships, covering the whole process of refined management from data measurement, cost accounting to fund collection and payment:

1. Accurate metering and data collection: through smart meters, sensors and communication networks, real-time collection of distributed energy power generation, power consumption by power users, charging and discharging amount of energy storage system and other key data, accurate to the second level or finer granularity, to ensure data integrity, accuracy, and to provide a basic basis for the settlement; at the same time, integration of meteorology, equipment operating conditions and other related information, to assist in the correction of power consumption and anomaly judgment, such as in the event of inclement weather affecting PV

power, according to the model Meanwhile, it integrates weather, equipment operating conditions and other related information to assist power correction and abnormality judgment.

2. Adaptation of multiple billing rules: customized billing scheme based on different transaction types, with flexible settlement of electric energy transactions based on time-sharing tariffs, ladder tariffs, or negotiated pricing, with full consideration of peak and valley hours, power consumption capacity, and power factor incentives and penalties; ancillary services based on the quality of contribution of FM, VAC, and standby capacity, and the length of time, and benchmarking of market rules to compute compensation costs, such as FM billing based on the speed of response, and the subdivision of the regulating precision; Distributed energy generation subsidies are accurately issued in accordance with regulations, distinguishing between different categories of photovoltaic, wind power, and waste heat, and calculating subsidies in proportion to feed-in and self-consumption, so as to ensure that revenues are compliant and reasonable.
3. Settlement process automation: the moment a transaction is concluded, the system automatically generates an electronic bill according to a preset template, clearly displaying the amount of electricity, unit price, cost details and settlement cycle, and pushes it to all parties online; integrating the financial software and payment gateway, and supporting multi-channels, such as UnionPay and third-party payment, to realize one-key payment and real-time payment, the whole process does not require manual intervention, greatly improving the efficiency of settlement, and the period of payment arrival has been reduced to 1–3 working days, reducing the cost of capital turnover. This greatly improves the settlement efficiency and reduces the payment arrival cycle to 1–3 working days, which reduces the cost of capital turnover.
4. Reconciliation and Dispute Handling: Regularly (daily, weekly, monthly) automatically compare the data of each metering point within the virtual power plant, and the data of the power grid and counterparties, and provide intelligent early warning and traceability to check the data deviation in the event of discrepancy; in the event of fee disputes, the blockchain deposit, contract terms and conditions can quickly trace back to the whole transaction, provide transparent and credible data vouchers, assist in negotiation and mediation, and built-in dispute arbitration mechanism, guaranteeing the settlement is just and free of ambiguities, and safeguarding the stability of cooperative relations. Stable cooperative relationship.

Financial statement and analysis: real-time generation of revenue, cost, profit and other core statements, visual presentation of the structure of transaction revenue, cost distribution, arrears situation; in-depth excavation of settlement data, analysis of business profitability, price sensitivity, customer credit risk, for the optimization of pricing strategy, market expansion, resource allocation adjustment output data insight, such as based on the analysis of the increase in the purchase price of the peak-time ancillary services to attract more resources to participate in the virtual power plant financial health and sustainable development of the drive. Drive the financial health and sustainable development of virtual power plants.

Aggregate Resource Management

Aggregate resource management for virtual power plants is the process of integrating, coordinating and optimizing various distributed energy sources and controllable loads, which is crucial for the efficient operation and value realization of virtual power plants. The following are its main functions:

1. Resource Access and Integration

    (1) Distributed energy access: Supporting multiple types of distributed energy access, such as photovoltaic, wind power, small hydropower, biomass, gas turbines, etc., realizing the interconnection and synergistic operation of different forms of energy, and providing diversified energy supply for the virtual power plant.

    (2) controllable load Integration: Integrate all kinds of controllable loads, such as electric vehicle charging piles, smart home appliances, energy storage systems, industrial users' uninterruptible loads, etc., and incorporate these decentralized load resources into the scope of management of the virtual power plant through intelligent control technology, so that they can be flexibly adjusted according to the demand of the power grid.

2. Resource Monitoring and Evaluation

    (1) Real-time data collection: With the help of advanced sensors and communication technologies, real-time data collection of distributed energy generation power, power, equipment status, etc., as well as information on power consumption, power, load characteristics, etc. of controllable loads is carried out to provide accurate data support for the operation and control of the virtual power plant.

    (2) Resource performance assessment: assess the performance of the accessed distributed energy sources and controllable loads, including power generation efficiency, reliability, adjustability, response speed, etc., in order to understand the actual operating status and potential of the resources, and to provide a basis for optimal allocation and scheduling of resources.

3. Optimized allocation of resources

    (1) Generation plan formulation: According to the grid demand, market price, generation characteristics of distributed energy resources and the regulation ability of controllable loads, a reasonable generation plan is formulated to realize the optimal combination and coordinated operation of distributed energy resources, and to improve the power generation efficiency and economy of the virtual power plant.

    (2) Load regulation strategy: For controllable loads, formulate flexible and diverse load regulation strategies, such as time-sharing tariff guidance, demand response incentives, etc., to encourage users to reduce the power

load during peak hours and increase the power load during low hours, so as to realize load shaving and filling in the valley, and to improve the stability and reliability of the power grid operation.

4. Resource Scheduling and Control

(1) Centralized scheduling management: establish a centralized scheduling management platform to carry out unified scheduling and control of distributed energy sources and controllable loads in the virtual power plant, realize real-time dynamic adjustment and optimal allocation of resources, and ensure that the virtual power plant can quickly respond to the grid's scheduling instructions and changes in market demand.
(2) Distributed cooperative control: Adopting distributed cooperative control technology to give full play to the autonomy and flexibility of distributed energy sources and controllable loads under the premise of guaranteeing the overall operation objectives of the virtual power plant, to realize the organic combination of local optimization and global optimization, and to improve the operation efficiency and reliability of the virtual power plant.

7.7.4 General Architecture for Function Realization

Intelligent energy management and control and virtual power plant platform is based on the IOT management platform, interaction integration, data center and other basic application capabilities to realize the interaction of equipment layer data and the application foundation of external systems, to support the upper level business management applications and to support the front-end business development; at the same time, it establishes a data-driven business innovation system internally in which the business and technological capabilities can be continuously deposited, and the products/services can be rapidly fused and innovated. The overall architecture diagram is as follows:

7.7.4.1 Overall Architecture Diagram

Below is Overall Architecture for Function Realizatiion (Fig. 7.6),

1. Business management applications mainly include: energy monitoring, energy audit, operation and maintenance management, carbon emission management, virtual power plant, configuration center and decision analysis display and other business functions.
2. Data collection and access: the IOT management platform realizes the interaction with the equipment layer data, and the platform supports a variety of ways, such as system-to-system docking and collection through the intelligent energy gateway,

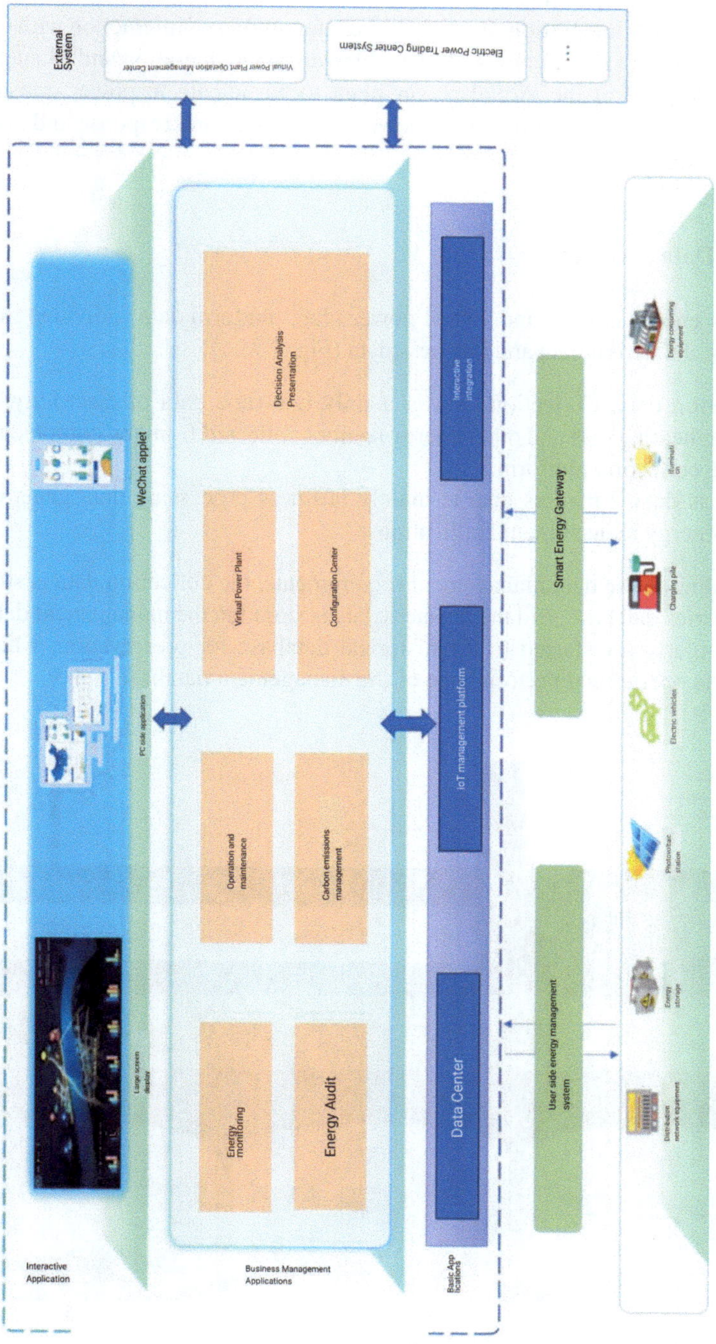

Fig. 7.6 Overall architecture for function realization

in order to adapt to the access needs of multiple types of heterogeneous energy systems and equipments.

3. Interaction and integration: It supports the integration and interaction with Guangdong Power Trading Center and Virtual Power Plant Operation and Management Center to support the virtual power plant to participate in power transactions such as spot power trading, green power trading, demand response and ancillary services.

7.7.4.2 Data Structure

Intelligent energy control and virtual power plant platform data is divided into two categories: collection data and business data (Fig. 7.7).

1. Collection data. Collection data is mainly real-time data of electricity, water, cooling/heating, gas and other energy sources collected from the equipment layer by IOT collection platform.
2. Business data. Business data is mainly business processing data generated by smart energy management applications.

According to the data management requirements, the collection data is stored in the time series database, and the business data is stored in the management database, and is regularly transferred to the historical database for preservation, which has ensured the operational performance of the management database.

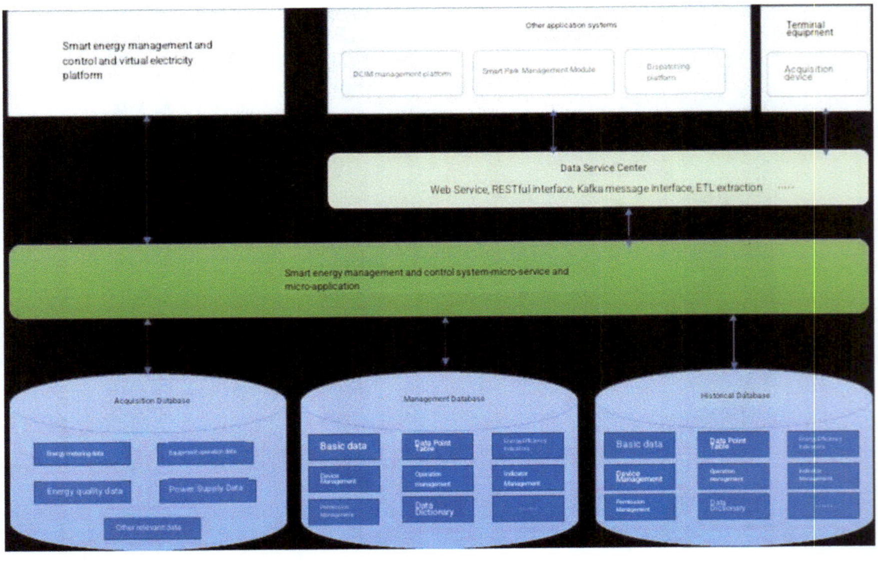

Fig. 7.7 Smart energy management and control system

7.7.4.3 Function Planning

Configuration Center

Business Configuration: Configuration of business parameters such as energy and carbon emission discounting system, energy price, peak and valley time configuration, energy efficiency knowledge base management, energy efficiency index library management, energy price management, etc. can be carried out.

System Configuration: Satisfy the basic configuration requirements for platform operation and business operation. This includes the configuration of basic system parameters and business operation parameters.

Interface Integration Design

Photovoltaic power generation collection and control integration: Integrate with the PV station system through the application to obtain the status of grid connection points, PV inverters, convergence boxes, box transformers, etc., climate and environmental information such as light, temperature and humidity, wind speed, etc., as well as power generation information.

Wind power collection and control integration: Integrate with the wind power station system through the application to obtain wind turbine operation status, wind power and wind power generation information.

Energy storage station acquisition and control integration: Integrate with the energy storage system of the energy storage station to obtain the status information of batteries, converters and other supporting auxiliary equipment, as well as charging and discharging operation information.

System Performance Index Design

Performance Design: Number of system measurement points Maximum 10,000 points; System response time General function response time <2 s, complex function response time <10 s; Real-time data collection cycle 5 min to 1 h free setting, initialization 15 min; System deployment mode Support virtualization.

Storage Design: Real-time data storage cycle 6 months; Historical data storage cycle 10 years; Database type Relational database to store business data and archive data, relational database storage should support SQL Server, MY SQL and other mainstream relational databases.

Maintainability Design

Configurable

Maintainability of personnel and organization: the system should have the function of maintaining basic information such as personnel/institution, and the system should be able to quickly maintain and adjust the operation of personnel/institution information.

Maintainability of post authority: the system should have the maintenance function of post authority, and the system should be able to quickly perform maintenance operations such as authority granting and recalling on post authority.

Maintainability

Maintainability of business processes: The main business processes of the system should have maintenance functions, and the system should be able to quickly adjust and maintain the business processes according to the changes in business rules.

Maintainability of service interfaces: The main business functions of the system should provide standard service exchange interfaces, which can be configured through switches to quickly provide external service capabilities.

Maintainability of parameter indexes: The system should have standardized and perfect management functions of parameter indexes, and have the functions of configuring and maintaining the basic performance parameters of system operation.

Monitorable

Provide log audit function: each component of the system should have standardized and perfect log management function, with multi-level log collection switch, valid/invalid switch, performance index collection switch and open configuration parameter table.

Standard monitoring protocol support: It meets the interface specification of the mainstream monitoring software in the industry, so that the monitoring data can be easily accessed to the monitoring software for centralized monitoring and management;

Readable and Easy to Modify

Requirements in the construction process of the system should be standardized, clear, complete and detailed documentation, such as business requirements stage to have business use case model, business rules, form certificate book, etc.; system requirements analysis stage requires a system use case model, use case documents, rule descriptions, etc.; the outline design stage requires a macro-design documents; the detailed design stage requires class diagrams, timing diagrams, etc.; the coding stage requires programming instructions, variable definition instructions, etc.; the testing stage requires a program design instructions The coding phase requires program design instructions, variable definition instructions, etc. The testing phase should have test cases, test records, etc.

Easy to Upgrade

Requirements for databases, application servers, development tools can be easily upgraded version, with downward compatibility; easy to upgrade also requires the client to upgrade the workload is small, the use of browser clients rather than GUI clients.

Human-Machine Interface Design

Easy to Understand

All the business functions of the system have the same interface style and operation flow; Business forms try to achieve WYSIWYG;

The interface is beautiful, simple and efficient; the layout of each component of the interface should maintain rationality and consistency;

Consistent interface style, color harmony, clear prompts, appropriate window size, easy to use; and In the choice of shortcuts, abbreviations, hints and icons should be in line with the user's habits.

Easy to Operate

Commonly used operations are supported by shortcut keys, and most of the operations can be completed within the keypad;

Information entry can be completed completely through the keyboard;

Both logical and operational steps should be avoided.

Easy to Learn

Provide online help, the key business operations of the system should provide online help documents and prompt information, so that operators can quickly and intuitively use this information to carry out the corresponding business operations, and provide timely feedback and prompts on various status and operation results.

Reliability Design

The correctness of business logic should be guaranteed under normal and extreme conditions;

The business system should meet the requirement of being available 24 h a day, 7 h a day;

System backup: provide backup system to prevent single point of failure.

7.7.5 Software Integration Architecture and User Interface Implementation

7.7.5.1 Software Integration Architecture

The virtual power plant operation platform integrates with the power trading platform, dispatching technology support system and new type of power load management system in the north direction, integrates with the power trading platform to realize the transaction declaration and the acquisition of clearing information, integrates with the new type of power load management system to realize the response of the demand response regulation plan and the uploading of data, and at the same time, through the security access area, it interacts with the dispatching technology support system, so as to realize the technical access and capacity detection.

At the same time, it carries out data interaction with the dispatching technical support system through the security access area to realize technical access and capacity detection, and carries out data interaction through the E file to realize the reception and parsing of dispatching instructions. The southward direction realizes the collection, monitoring and dispatching control of adjustable resources at the regional and station levels through communication with the edge intelligent gateway.

Below diagram is for Software Integration Architecture (Fig. 7.8).

Software Data Following

1. Real-time data collection and integration
   Intelligent sensors are deployed in all aspects of the production workshop to accurately collect key parameters such as voltage, current, temperature of electrolytic aluminum tanks, and real-time power of various power equipment and lighting systems, which are matched with high-speed communication network to realize millisecond data transmission back to the central control system of the virtual power plant. At the same time, it deeply connects with the production management system (MES), obtains information such as production scheduling plan, product batch process requirements, etc., and integrates them to form a real-time database of load panorama, which builds up a solid data foundation for accurate regulation.
2. Load characteristic modeling and analysis
   Based on the massive load data accumulated over a long period of time, machine learning algorithms (e.g., cluster analysis, neural network) are used to analyze the load curve of the aluminum industry, accurately identify the changing pattern of load under different production processes and time periods, and construct a multi-dimensional load characteristic model. Distinguish between the rigid base load, auxiliary processes can be flexible control load, quantify the load of each link on the price of electricity, grid frequency fluctuation sensitivity, for the generation of refined control strategy to provide a scientific basis.
3. Dynamic regulation strategy generation

   (1) Following the time-sharing tariff mechanism, during the valley power hours (e.g., late at night to early in the morning), the virtual power plant system automatically issues commands to moderately increase the adjustable current, accelerate the process, and increase the production without affecting the quality of the product; during the peak period (daytime peak), the load of non-critical equipments (e.g., part of the ventilation and material transfer equipments) will be cut in an orderly manner according to the preset priority, or the electrode parameters of electrolysis tanks can be fine-tuned, and the overall power will be reduced in an even manner. 5–15% to ensure the optimal cost of electricity consumption.

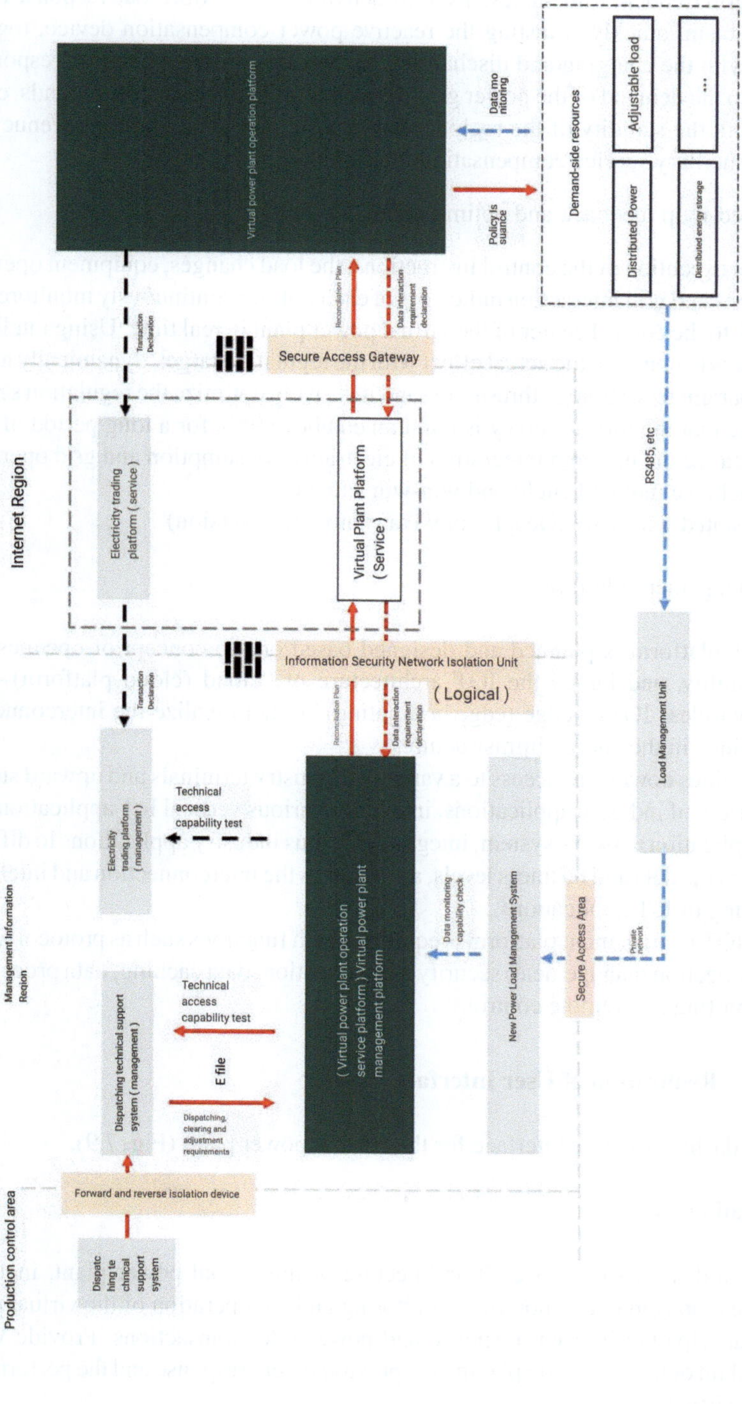

Fig. 7.8 Software integration architecture

(2) Closely tracking the real-time operating status of the power grid, when the grid frequency fluctuates, instantly activating the flexible load response mechanism, quickly adjusting the reactive power compensation device, together with the charging and discharging of the energy storage facility, responding to the demand of the power grid frequency adjustment in milliseconds, ensuring the stability of the regional power grid, and obtaining the revenue from ancillary service compensation through this.

4. Closed-loop feedback and optimization

After the execution of the control instructions, the load changes, equipment operating conditions and grid interaction indicators of each link are continuously monitored and fed back to the control center of the virtual power plant in real time. Using intelligent algorithms to compare the actual effect with the regulation target, dynamically amend model parameters, strategy thresholds, and iteratively optimize the regulation scheme to ensure that the load accuracy is maintained above 98% for a long period of time, and to realize the in-depth integration of electricity consumption and grid operation, and to achieve mutual benefit and win-win situation.

Translated with www.DeepL.com/Translator (free version)

IOT Management Platform

The IOT platform is planned and designed based on the concept of openness and compatibility, and builds the IOT architecture of "cloud (cloud platform)—pipe (wired/wireless IOT)—edge (edge computing)" to help realize the interconnection of all things in the energy infrastructure.

It provides downward access to a variety of industry terminals and upward support for a variety of industry applications, integrates various vertical IoT applications into a flat application network system, integrates various industry applications to different degrees at the data and business levels, and realizes the interconnection and intelligent processing of IoT applications.

The IOT management platform is equipped with functions such as protocol resolution, connection management, security authentication, data caching, data processing, data reporting and remote control.

7.7.5.2 Realization of User Interface

Here is dashboard User Interface for this virtual power plant (Fig. 7.9),

Aggregation UI

Provide a display of the overall architecture of the virtual power plant, including resource management, panoramic monitoring and the operation of the virtual power plant participating in event response and power sales transactions. Provide virtual power plant output curve, displaying the previous event response and the performance of each virtual member.

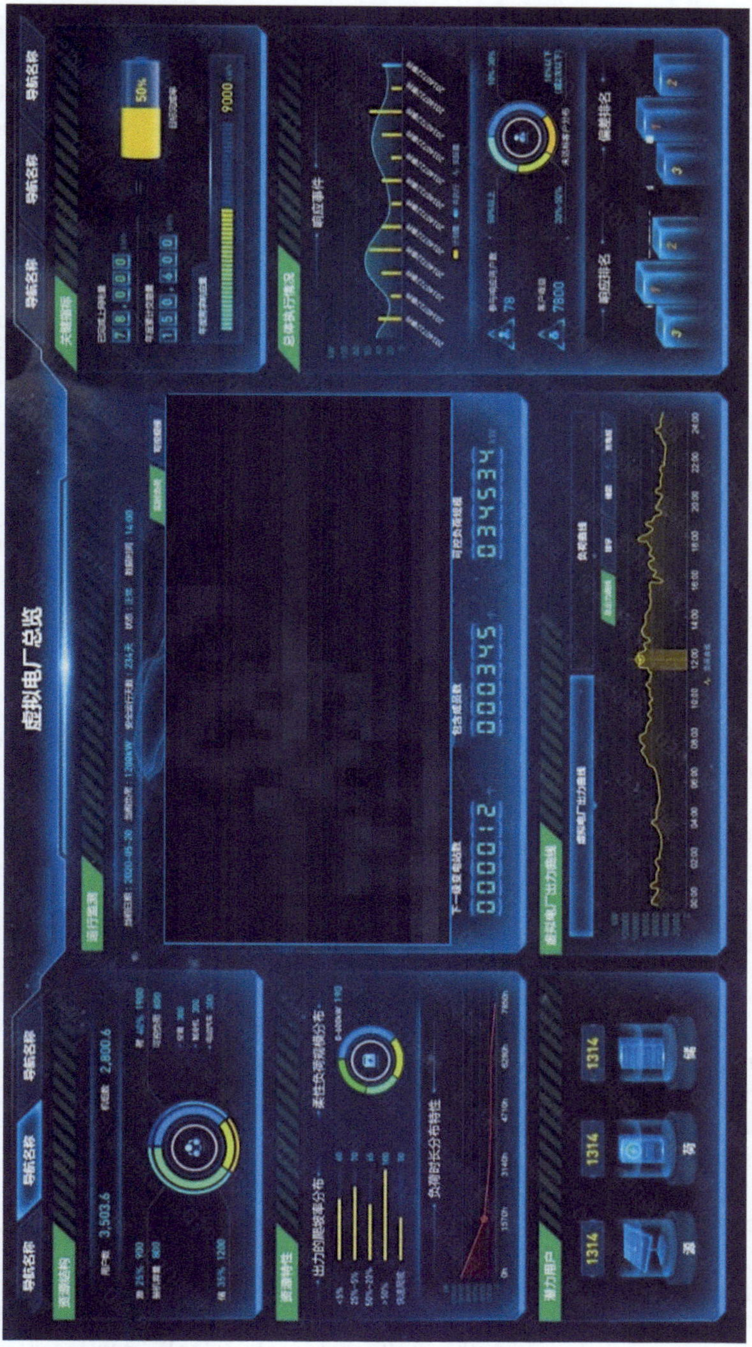

Fig. 7.9 Dashboard user interface

Virtual Power Plant Aggregation Resources

Demonstrate the number of various resource types, resource characteristics (creepage rate, response time, etc.) and adjustable status display. Utilizes marketing data (load, current, voltage, etc.), conducts feature mining for identification, and utilizes customer profiling to obtain information on the customer's industry type, the power supply sub-district in which it is located, etc., and analyzes the size of its flexible resources.

Virtual Power Plant Operation and Control Platform Monitoring

Show the distribution of each distributed virtual power plant resource in each transformer site in the form of a map, and provide a display of market analysis indicators; a display of virtual power plant intelligent management and control platform indicators; and an overview of platform operation.

Virtual Power Plant Business Operation Overview

Provides virtual power plant aggregation performance display; real-time output power curve of resources; clearing price display function; historical total service query function.

Monitoring UI

The functions realized by the virtual power plant intelligent management and control platform include:

Operation Monitoring

The main responsibilities of the energy operation monitoring module in the platform include real-time data acquisition and processing of aggregated resources such as photovoltaic, energy storage, charging piles, air conditioners, etc., energy status estimation, performance analysis, abnormality detection and alarm. These functions enable the virtual power plant to respond flexibly to grid demand and market changes, while ensuring equipment performance and system security. Through operation monitoring, it is able to provide users with digital monitoring, drive the digital transformation of upstream and downstream SMEs, and provide certain basic business income for the virtual power plant.

Photovoltaic Monitoring

Utilizing energy Internet technology to manage a large number of dispersed PV power stations, it conducts power generation analysis, operation environment monitoring, discrete rate analysis and other multi-angle analysis for PV power generation.

Energy Storage Monitoring

Monitor the performance and status of energy storage equipment, collect and analyze key parameters in real time to ensure the safe and efficient operation of energy storage equipment, such as: remaining power, charging load, discharging load, and health condition.

Charging Pile Monitoring

Monitor the working status and electrical parameters of charging piles, such as: charging pile status, charging load, standby load, and order information.

Industrial Load Monitoring

Monitor enterprise voltage, current, power, power factor, electricity and other data

Air conditioning monitoring

Monitor the operation status, energy consumption, temperature and other data of air conditioners through intelligent technical means.
Lighting Monitoring

Monitor the operation status, energy consumption, illumination and other data of lighting through intelligent technical means.
Data Center Monitoring

Ensure the efficient, safe and reliable operation of data centers, monitor the power load of data centers in real time, and ensure the stable operation of data centers.
User Profile and Metering Point Management

Maintenance, viewing, addition and modification of user basic files, PV, energy storage, charging pile, air conditioning, lighting, industrial equipment and other resource files. Maintenance and management of monitoring points, metering points and other information

Aggregate Resource Management UI

After accessing the resources, the unified management of the resources is realized, which mainly includes the management of aggregated resource files, user file management, metering point file management, management of contract attachments, and management of documents reported to the power trading center.

Virtual Unit Management

Maintains and manages each unit under the virtual power plant according to the unit, user, and circuit resource maintenance. It also includes changes, cancelations, and approvals of units.

Resource Evaluation Management

According to the ability of access resources, assess the priority, and provide support for the subsequent demand response and electricity energy trading to do marketization.

Aggregation Monitoring

Overall monitoring of the virtual power plant resource pool and monitoring of the unit dimension, statistics and advanced display through visualization charts, support for linkage and interaction with curves, details and other functions.

Regional Monitoring

Monitor the accessed aggregated resources by region, realizing all-round monitoring of the aggregated resources under the stratification and grading of resources.

User Monitoring

Through real-time collection and analysis of power consumption data on the user side, including power consumption, power consumption patterns and other information, it realizes an accurate grasp of the user's power consumption behavior.

Split-circuit monitoring

Focusing on the power output of each sub-circuit of users under the virtual power plant units, through the monitoring and management capacity building of aggregated sub-circuit resources, it realizes the tracking of the state of the controllable resources of the virtual power plant and the assessment of the adjustable capacity, so as to provide a lean data base of adjustable capacity for the participation of the virtual power plant in the market.

Equipment Monitoring

Real-time collection and display of equipment operating status, performance data and key parameters through each metering point. Understand the current status of equipment and take appropriate measures when needed.

Statistical and Analysis UI

Energy Efficiency Analysis Module

The platform system analyzes and processes the year-on-year and year-on-year energy consumption of each category and sub-category, as well as different time phases according to the summarized and statistical energy consumption data, which can intuitively show the different impacts of different seasons and different working modes on energy consumption, formulate a reasonable and effective energy consumption strategy based on the results of the analysis, and tap into the energy-saving potentials of different equipments. It helps users to realize energy consumption analysis and find solutions to manage energy saving, which is mainly used in digital transformation, energy-saving transformation, power distribution operation and maintenance and other business models.

Energy Consumption Statistics

Shows the overall energy consumption status by region, time period, network power supply, green power, and carbon dimensions. As below diagram shows (Fig. 7.10).

Energy Consumption Statistics Interface

Regional Energy Consumption Analysis

By analyzing and ranking the energy consumption of different regions, it can be used to monitor the regional energy consumption, compare the energy consumption of the same period and the same ring comparison.

Regional Energy Consumption Analysis Interface

Time Period Power Consumption Analysis

Time period power consumption analysis shows the power consumption and tariffs in each time period of peak and valley, and the proportion of power consumption and tariffs in each time period. By observing the power consumption and tariffs in different dimensions and regions, it can detect abnormal power consumption in time, and provide data support for guiding the users to adjust their own power consumption habits and save their power tariffs.

Time Period Power Consumption Analysis Interface

Equipment Energy Consumption Analysis

Equipment energy consumption analysis is to classify, analyze and manage the energy consumption of equipment in a fine-grained manner, and to analyze the energy consumption of equipment for different types of equipment, key equipment,

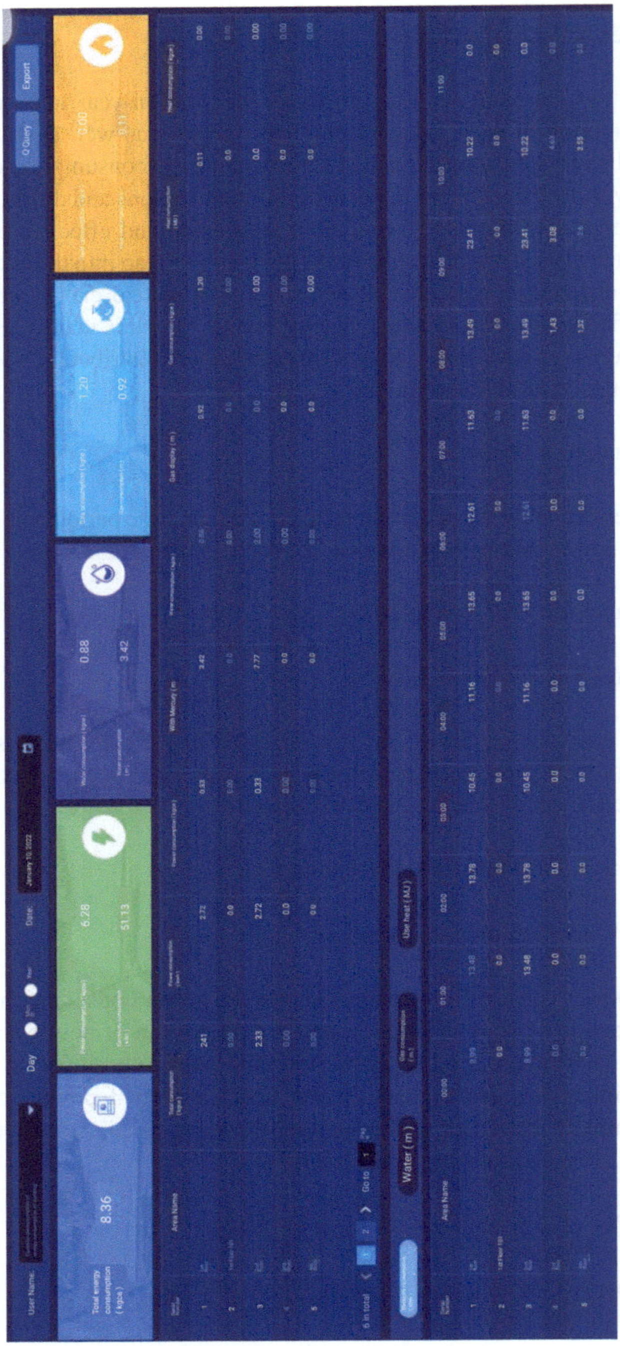

Fig. 7.10 Overall energy consumption status

the same equipment at different times, the same type of equipment, and other situations, as well as to make horizontal comparisons of the energy consumption of equipment and vertical year-on-year comparisons and ring comparisons of the equipment's own energy consumption, and to visualize the data display through charts and graphs, so as to dig out the energy-saving potentials of the equipment.

Report and Log UI

Equipment Energy Consumption Analysis Interface

Energy Consumption Analysis Report

According to the relevant standards, the energy consumption analysis report can automatically generate the relevant energy consumption analysis report of the whole building, and the energy consumption analysis report can be exported to PDF, so as to systematically analyze and display the whole building through the energy consumption analysis report.

Energy consumption analysis report interface figure.

Scheduling and Strategies UI

Cooperative Scheduling Module

Receive northward scheduling data, realize southward internal resource scheduling allocation, realize electric energy market transaction through operation strategy, and support electric energy transaction.

Energy Storage Equipment Strategy Management

Monitor the current strategy implemented by the energy storage equipment and remotely modify the energy storage charging and discharging strategy.

Operation Strategy Management

Scheduling operation strategy management, scheduling and allocation strategy for resources in the virtual unit.

Scheduling Operation Monitoring

Monitor the operation of resources within the virtual unit, including status, load, voltage, current and other data.

Scheduling Log Monitoring

Used to collect, store, and view northbound scheduling commands to realize traceability.

Timed Task Management

Manages the time of scheduling operation through timed tasks.

Transaction Aid Decision Module

The trading assistant decision module is mainly used to assist managers in making market trading decisions. Through data collection and processing, it carries out market analysis, formulates trading strategies, and provides decision-making suggestions to help managers declare at the right time with reasonable power and reasonable price, and to help operators participate in electric energy trading.

Load Forecasting

Forecasting the power supply capacity on the supply side and the electricity demand on the demand side by collecting data on historical market supply and demand, power forecasts, etc.

Generation Forecast

Forecasting power generation in the electricity market based on relevant operation information of the power generation side, historical power generation information and regional policy dynamics.

Market Price Forecast

Based on the influencing factors of electricity price and the historical data of electricity price in the market, we use relevant models to forecast the spot electricity price, medium and long-term electricity price, and nodal electricity price.

Transaction Declaration Strategy

Based on market forecasts and contract data, provide comprehensive support for virtual power plants to participate in trading strategies, such as trading time, trading power, trading offer, etc.

Power Generation and Consumption Plan Management

Used to participate in market-based trading, to realize the preparation and management of power generation plan.

Operation and Revenue Analysis

Analyze the revenue trend of distributed resources and predict the return cycle through multi-dimensional power and revenue statistics.

Demand Response UI

Demand Response Module realizes comprehensive, accurate and fast response to market demand or customer demand by integrating data analysis, forecasting and response technologies to help operators participate in the demand response market.

Baseline Analysis

Baseline analysis is mainly responsible for collecting and analyzing historical data to establish a demand baseline. It collects user baseline load, maintains and manages user baseline, and realizes unit baseline analysis management and power plant baseline analysis management according to unit and power plant maintenance.

Response Strategy

Through resource evaluation and response strategy, the selection of resources within the virtual unit is realized, and when the demand response index is confirmed, the resource with the best scheduling and economy is selected to participate in the demand response.

Demand Response Records

Manage current, historical and upcoming demand responses, and maintain and manage all demand responses.

Response Information Maintenance

Maintains the data declared for demand response, including the first grade, second grade, peak shaving, valley filling, and so on.

Response Capacity Confirmation

Manual confirmation of the demand response plan, determination of the invited capacity of the aggregator issued by the provincial platform, formulation of the actual execution plan of the whole system, construction of the pool of aggregation and regulation resources of adjustable resources, confirmation of the algorithm of the multi-layer and multi-dimensional aggregation strategy for calculating the execution parameters of the selected resources, and adjustment of the variable parameters within the range, to realize the flexible regulation and rigid control of the various types of access adjustable resources.

Response Plan Issuance

The platform receives demand response invitations, confirms the response information and formulates the plan, automatically generates response scheduling strategies according to the invitation target, and issues corresponding control instructions to the response resources.

Response Process Monitoring

After the response program is formulated and issued, it will be executed and monitored. During the execution phase, it is necessary to ensure that the measures and strategies are carried out in an orderly manner according to the predetermined plan. At the same time, the monitoring of power system operation status, timely detection and resolution of possible problems and risks, to ensure the stability and reliability of power supply.

Response Effect Evaluation

User participation in demand response carries out demand response strategy effect evaluation in different scenarios, and analyzes the results of system participation in demand response, including response completion, response gain analysis, etc. Evaluation of response regulation operation is carried out through economic gain analysis and deviation calculation, including regulation response time, regulation deviation, response completion rate, etc. Meanwhile, it provides evaluation reports and suggestions to help enterprises summarize lessons learned and continuously improve and optimize demand response strategies.

Strategy Control

According to the real-time demand of the power system and the scheduling instructions of the power grid, the output of distributed energy resources is automatically adjusted to meet the balanced demand of the power system. Through demand response, the virtual power plant system can effectively cope with fluctuations and uncertainties in the power market. It is responsible for the trading of electric energy between the virtual power plant system and the power market, automatically formulating trading strategies and executing trading operations according to the real-time electricity price and supply/demand situation in the power market, so as to realize the optimal allocation of electric energy and the maximization of economic benefits.

Benefit Assessment

Through data analysis and visualization technology, it demonstrates the green and low emissions of the virtual power plant system in terms of energy saving, emission reduction and carbon emission reduction.

7.7.5.3 Intelligent Transaction and Control Console

Resource Forecast Analysis

(1) Load Forecasting
Medium- and long-term load forecasting: Based on the geographical, meteorological, type and capacity information of single-point and cluster loads, the load template is carved to realize the medium- and long-term load curve forecasting on monthly and yearly basis; Short-term load forecasting: Based on the historical data of power and meteorology of single-point and cluster loads, it provides forecasting methods based on time series, regression analysis and artificial neural network to realize minute and hourly short-term load curve forecasting.
(2) Power Generation Forecasting
Medium- and long-term new energy power generation forecast: Based on the geography, meteorology, type and capacity information of single-point and cluster new energy sources, it carves out regional new energy processing templates and realizes monthly and annual medium- and long-term new energy power generation capacity forecast; Short-term new energy power generation forecast: based on the historical data of power and meteorology of single-point and cluster new energy sources, it provides forecasting methods based on time series, regression analysis and artificial neural network, realizing short-term new energy power generation capacity forecast in minutes and hours.

Dispatch Potential Analysis

Based on massive operation data, combine resource type, resource attribute, resource adjustability, resource adjustment potential, load characteristic analysis, and carry out the construction of the park's source, network, load and storage all-resource element model. A multi-dimensional evaluation index system including response speed, response reliability, etc. is established, and an evaluation model considering the distribution of scheduling capacity in time scale is built with time sequence constraints. In addition, for different normal/emergency operation scenarios, data dimensionality reduction, correlation analysis and other methods are used to analyze the minimum set of input parameters of the evaluation model, and propose a multi-timescale online evaluation model of the scheduling capacity of different aggregate resources of the virtual power plant. Evaluation model. Based on the regulation potential analysis model, the regulation potential of different aggregation units is obtained, including whether they are adjustable or not, the adjustable time period, the maximum output power, the minimum output power, the maximum number of equipments, the minimum number of equipments in different time periods, and so on. Get a clear picture of their response potential and response periods, etc.

Market Transaction Declaration

It supports the business management of virtual power plant market-oriented trading operation before, during and after the event, and provides technical support for the trading process and planning functions such as program preparation, trading declaration and trading results.

1. Transaction announcement management. Provide management functions for transaction announcement information and transaction constraint information. Provide the maintenance function of transaction results; the transaction results can be viewed by power plant operators.
2. Transaction program management. According to the transaction announcement information, the new transaction declaration program can be added, and can be modified and deleted before the program is audited; the program cannot be adjusted after the program is audited; the program includes the load curve, time period and resource declaration plan, and when editing the transaction declaration program, the distributed resource plan in the program can be added or deleted according to the needs.
3. Market declaration management. On the basis of load forecasting, the organization of virtual power plant resources to marketing real-time power demand side management system, trading platform and other related platforms for market declaration.
4. Market declaration model. Provide strategies for participation in the power market declaration to meet grid scheduling and market demand, as well as to maximize the benefits of the aggregation unit.
5. Plan issuance. Import, display and query the grid-issued plan curve or transaction clearing results.

Intelligent Regulation Management

According to charging load forecast, system scheduling demand, power market information and other information, taking into account factors such as response range, rate, time period, etc., study the intelligent regulation and control strategy of virtual power plant resources participating in the operation of the power grid, strengthen the information gathering, data analysis and information flow capability of the load aggregation application, construct the multi-time scale coordinated control model of pre-date optimization—intra-day rolling—real-time balancing, and achieve the intelligent decision-making of the interaction between the virtual power plant and the grid, so as to satisfy the power grid's needs and maximize the benefits of the aggregation unit. Intelligent decision-making to meet grid scheduling needs and maximize user benefits.

Real-Time Interactive Monitoring

Fully display the results of regulation and control of various types of load resources of the virtual power plant, including the grid issued plan, control baseline, real-time power, regulation volume, gap volume and other information, and realize the visualization of the display. Monitor the execution process of response events, support querying the execution process monitoring situation of the executing party (planned regulation load and actual regulation load) by date and by event.

Effectiveness Evaluation

Obtain the effect of virtual power plant resources participating in response execution on the provincial side, support comparison and analysis between the monitoring data by platform and the execution effect obtained by the provincial test, check the deviation of response, and support feedback of objections to the execution effect.

Subsidy Benefits

Obtain information on the subsidy gain from the participation of virtual power plant resources in response on the provincial side, and support multi-dimensional display of the gain from the participation of virtual power plant resources in power market transactions.

Decision Analysis Display

Decision analysis display as a 7 * 24-h uninterrupted monitoring, to achieve full-time management of energy use, energy supply and demand, carbon emissions analysis within the data center, to provide real-time data of the entire data center campus for the duty officer, to do a good job of energy management and equipment management, dealing with emergency accidents to provide first-hand information.

The overall display style of the big screen needs to be technology-oriented, and through visual dynamic effects and dynamic numerical index changes, it can realize dynamic monitoring and analysis of the energy consumption of the park's data center.

The construction of the big screen needs to include four major themes: energy consumption management, energy efficiency management, carbon emission management and operation and maintenance center:

Energy Consumption Management

Realize real-time monitoring of data center real-time electricity, heat, water, gas, cold and other energy sources. Through digital twin technology, it realizes real-time dynamic monitoring of data center building building and IT equipment, air conditioning HVAC system, photovoltaic system, energy storage system, charging facility system and other kinds of energy systems, and supports viewing real-time operation data of equipment. It realizes the analysis and comparison of data center energy consumption by day, month, quarter, year and other multi-dimensional perspectives, and the energy consumption indicators include comprehensive energy consumption, per capita energy consumption, energy consumption per unit area and so on.

Energy Efficiency Management

Calculate the energy efficiency indexes of data centers and buildings in the park from multi-dimensional perspectives such as daily, monthly, quarterly and yearly, analyze the energy consumption indexes against the standard, and display them in the form of charts and graphs. This includes statistical benchmarking analysis of key energy efficiency indicators such as data center power utilization efficiency (PUE), water utilization efficiency (WUE), local power utilization efficiency (pPUE), cooling load factor (CLF), power supply and distribution load factor (PLF), renewable energy utilization (RER) and so on.

Carbon Emission Management

Carbon emission of the park is dynamically monitored and analyzed, and carbon emission of the park is compared and analyzed in different cycles, such as daily, monthly, quarterly and yearly. Carbon emissions of buildings, server rooms and equipment in the park are analyzed vertically and horizontally, and optimization suggestions are given to high carbon emission objects. Support to view carbon balance routes, and view the recommended carbon balance planning.

Operation and Maintenance Center

Display real-time alarm center content, statistics and dynamic analysis results on the number of alarms by alarm level and alarm status. View the location of alarm points and alarm details in the campus buildings and data centers in real time. Support viewing inspection logs and inspection results, maintenance and repair results. Display the results of activated and closed work orders in the work order center, and view the statistics and evaluation results of work orders.

References

1. Yao, F., Xie, G., Li, C., Liu, X., Ren, B., Lu, H., Li, R., Song, W., Duan, C., & Wang, W. (2024). Flexible construction technology and application of new integrated energy systems. In *Proceedings of the 2024 new power system development forum* (pp. 94–99). China Electric Power Press.
2. Yu, J., Xu, J. J., Yao, F., et al. (2024). *Integrated smart energy technology*. Beijing: China Electric Power Press.
3. Cheng, H. Z., Hu, X., Wang, L., et al. (2019). Review on planning of regional integrated energy systems. *Automation of Electric Power Systems, 43*(07), 2–13.
4. Zhong, Y. J., Ji, L., Li, J. X., et al. (2022). Overview of basic characteristics, connotations and development status of virtual power plants. *Integrated Smart Energy, 44*(06), 25–36.
5. Yao, F. (2023). Engineering on the basis of big data in virtual power plant. In *Proceedings of 24th conference on electric power supply industry, Xiamen, China.*

Chapter 8
Financial Analysis

Abstract This chapter focuses on the financial analysis of a virtual power plant project. The construction cycle, covering planning, development, construction, and operation stages. Major equipment and materials for data collection, networking, and system operation are detailed, along with software resources. The virtual power plant's business model yields diverse benefits. Demand-side response offers subsidy income. The ancillary service market provides revenue from frequency modulation, voltage regulation, and standby services. Spot market trading brings gains from electricity sales and cost savings. Energy saving and emission reduction result in carbon trading and environmental subsidy revenues, while capacity compensation reduces expenditure and earns compensation. Overall, building a virtual power plant optimizes energy utilization, cuts electricity costs, stabilizes the grid, and promotes sustainable energy practices.

8.1 Brief

Driven by the wave of digital transformation of energy and the reform of electricity marketization, the development project of basic virtual power plant has become a key puzzle piece in the construction of a new power system with its unique resource integration and collaborative dispatch capabilities.

The economic evaluation system described in this chapter provides a comprehensive framework and methodology for schedule planning, equipment and material statistics and initial investment estimation, revenue prediction modeling, and quantitative risk analysis to deeply analyze the project's economic feasibility and potential value space under the complex market environment, and to provide accurate quantitative basis and forward-looking guidance for scientific decision-making, rational resource allocation, and long term operation of the project.

F. Yao, *Engineering Guide to Planning and Design of Virtual Power Plant*,
https://doi.org/10.1007/978-981-95-2231-6_8

8.2 Methodology for Scheduling of Virtual Power Plant Project

The project schedule for the construction of the virtual power plant project is divided into several key phases, including planning, development, construction, and operation. The construction schedule and project's duration are keystone for economic evaluation of the project's time-value in serial and financial budget estimation.

One detailed schedule of the project is as follows:

Category	Content	April	May	June	July	August
Planning phase	Construction plan preparation					
Planning phase	Requirements research					
Planning phase	Prototype design					
Customization&Implementation	IoT data collection software					
Customization and implementation	Operation monitoring					
Customization and implementation	Aggregated resource management					
Customization and implementation	Energy efficiency analysis module					
Customization and implementation	Collaborative scheduling module					
Customization and implementation	Transaction decision support module					
Customization and implementation	Demand response module					
Customization and implementation	Large screen display					
Customization and implementation	Factory testing					
Construction and deployment	Equipment procurement					
Construction and deployment	Enterprise refurbishment and data access					
Construction and deployment	System integration, hardware debugging, system installation, system joint debugging					
Construction and deployment	Leading new energy data access					
Operation and acceptance	Trial operation					
Operation and acceptance	Acceptance					

This detailed schedule outlines the sequential workflow for a virtual power plant project across five months (April to August), structured into distinct phases. It begins with the Planning phase, which involves three key tasks: preparing the construction plan, conducting requirements research to understand project needs, and creating a prototype design to shape initial concepts.

Next is the Customization and implementation phase, a comprehensive stage focused on developing and refining core software components. This includes building IoT data collection software for gathering energy-related data, setting up operation monitoring systems, managing aggregated resources, analyzing energy efficiency, enabling collaborative scheduling, providing transaction decision support, implementing demand response capabilities, designing large-screen displays for data visualization, and conducting factory testing to ensure software functionality.

Following that is the Construction and deployment phase, which centers on physical and system-level setup. Tasks here range from procuring necessary equipment, refurbishing enterprises and establishing data access, to integrating systems, debugging hardware, installing systems, performing joint system debugging, and connecting leading new energy data sources.

Finally, the project concludes with the Operation and acceptance phase, where the system undergoes trial operation to identify and resolve any issues, and then formal acceptance to confirm that all project objectives have been met. Each phase and its associated tasks build on the previous one, guiding the project from initial planning to full operation and validation.

By preparing one detailed schedule, a structured framework for the construction of the virtual power plant project outlines the key phases, tasks, and expected timeline. The schedule serves as a roadmap for project management, resource allocation, and financial evaluation, ensuring that the project is executed with precision and efficiency. Thus, it is one fundamental aspect in financial analysis.

8.3 Equipment and Material Statistics

8.3.1 Material and Equipment

The quantity of materials for the construction of a virtual power plant project is calculated. Here is the architecture for a virtual power plant (VPP) (Fig. 8.1).

For each user (User 1, User 2, User 3, etc.), smart meters—connected to LoRa devices via Modbus—collect energy-related data (consumption, generation, etc.). These LoRa devices form localized LoRa clusters, which communicate with 4G gateways. The gateways leverage the 4G network to transmit the aggregated data to a cloud server. This infrastructure enables the VPP to remotely gather, monitor, and manage energy information from distributed resources, facilitating real-time insights, optimized resource allocation, and coordinated operation of the virtual power plant's energy assets.

Meanwhile, loop monitoring is realized by adding transformers and meters to the circuits of controllable equipment, and at the same time accessing PV, energy storage, and charging pile resources, and the number of related users is 10. For this VPP project, counting the number of user for the hardware equipment to be used (Table 8.1).

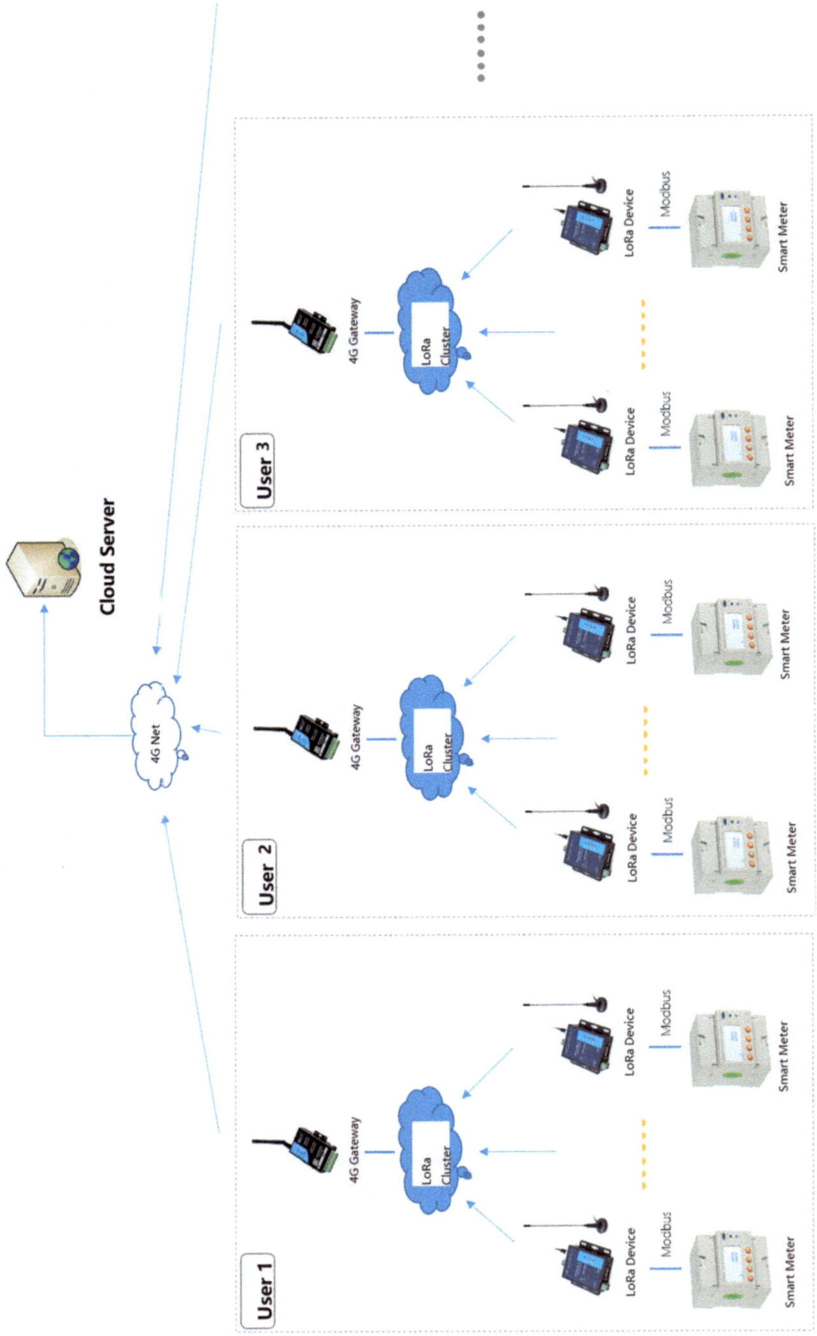

Fig. 8.1 One VPP project structure

Table 8.1 Material list for loop monitoring equipment

No.	Content	Unit	Quantity
1	System integration and commissioning	Unit	1
2	Bus and circuit data access	Unit	10
3	PV station data access	Unit	3
4	Energy storage station data access	Unit	1
5	Charging station data access	Unit	1

Among them, PV site monitoring is provided by the inverter of the site, energy storage is provided by the station-level ems system, charging pile is provided by the charging platform, monitoring data and interface are provided by the PV, energy storage and charging pile builders, and no hardware equipment is involved. 10 enterprise loop monitoring involves transformers, multi-function meters, LORA and other equipment, and the detailed hardware list is as follows (Table 8.2).

Server procurement

For deployment, cloud servers are adopted, mainly using servers and cloud security centers as the primary equipment. This setup ensures that data related to the virtual power plant, such as energy consumption information from smart meters and operation data of distributed resources, can be securely stored and processed. The servers handle the massive data flow, while the cloud security centers safeguard the integrity and confidentiality of the data, protecting it from unauthorized access, cyber-attacks, and ensuring the overall security of the virtual power plant's operation and management system (Fig. 8.2).

Services are deployed in clusters to ensure system stability. For example, Front End Servers, Business Servers, Acquisition servers all use cluster-like structures (Web server with load balancing, Application server, Collect clusters). The Database servers adopt a primary-standby setup for data backup, with relational databases having one master database for main operations and one slave for replication, ensuring data security and system reliability. Users access via devices connected through switchboard and Security Center, and equipment data is collected via LORA and 4G gateway into the system for processing like data acquisition and protocol parsing in different server layers.

Here is the Server procurement list (Table 8.3).

List of software resources

Table 8.2 Material list for detail hardwares

No.	Name	Model/Specification	Unit	Quantity	No. of users	Sub-total
1	Multi-function meter	RS485 Remote Communication Port to provide 3-phase current, 3-phase voltage, 3-phase active power, 3-phase reactive power, 3-phase apparent power, Total power factor, Forward active power, Reverse active power, Forward reactive power, Reverse reactive power and other parameters. Rail-mounted backplane with matching snap-on transformers	Unit	10	10	100
2	LORA device	Support point-to-point transmission, uplink LORA protocol, downlink modbus RTU protocol. 6000 m @ 0.268 Kbps support LBT detection, forward error correction, data retransmission, timeout no data restart	Unit	10	10	100
3	4G networking equipment	Industrial 4G router, support modbus RTU/TCP protocol conversion and modbus/Json protocol data reporting, including 2 years 4G traffic	Unit	1	10	10
4	Accessories and installation	Equipment installation, accessories and accessories, transportation, food and lodging, and other implementation costs	Unit	1	10	10

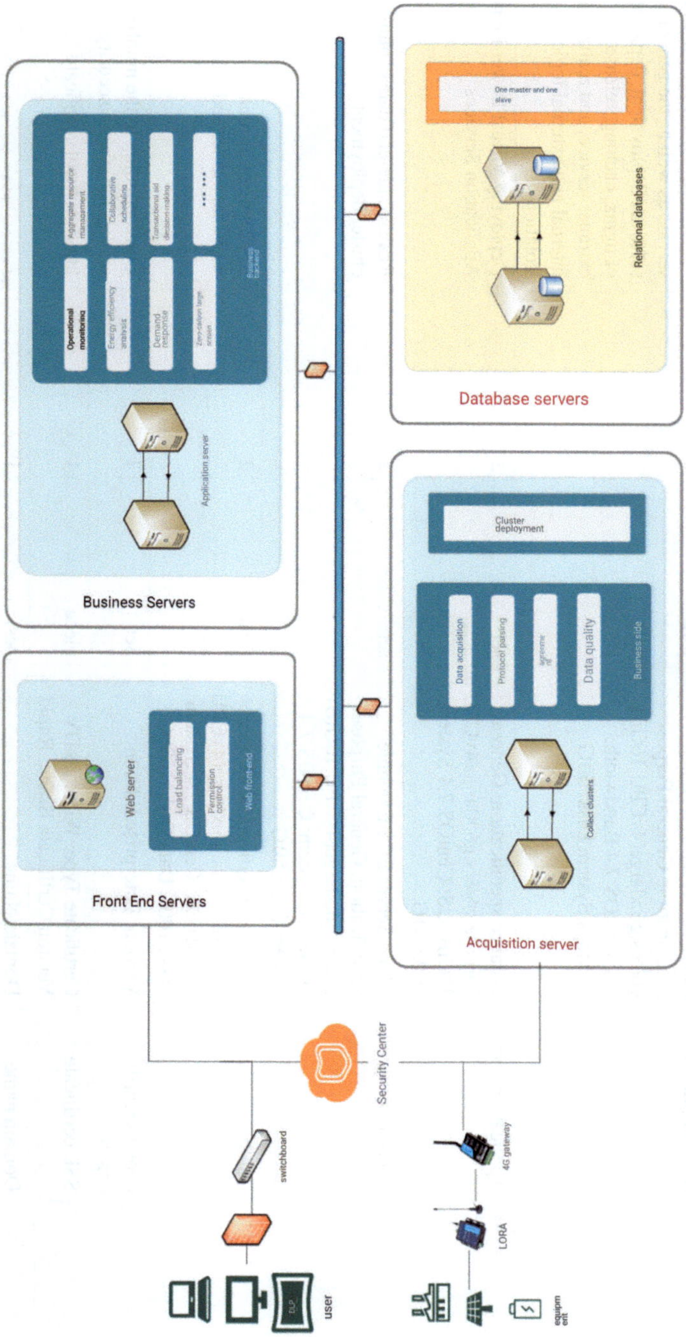

Fig. 8.2 VPP server procurement configuration

Table 8.3 Server procurement list

Serial No.	Name	Specification	Unit	Quantity	Remarks
1	Basic application	Cloud server ECS, Instance specification: General Purpose g6/ecs.g6.xlarge(4vCPU 16GiB) OS: CentOS 7.4 Bandwidth: 30 Mbps System Disk: 50 G	Unit	1	Mainly responsible for the extranet service as well as the general software deployment, deployment of nginx, caching, release of extranet service, and realization of internal and external logical isolation
2	Application server	Instance specification: General Purpose g6/ecs.g6.xlarge(4vCPU 32GiB) OS: CentOS 7.4 System Disk: 50 G	set	2	Deploying Virtual Power Plant Application Services
3	Capture server	Cloud Server ECS Example Specification: General Purpose g6/ecs.g6.xlarge(4vCPU 16GiB) Operating System: CentOS 7.4 System Disk: 50 G	set	2	Realize side-end data capture with cluster deployment
4	Database server	Instance Specification: General Purpose g6/ecs.g6.xlarge(4vCPU 16GiB) OS: CentOS 7.4 System Disk: 100G Data Disk: 1000G	Desk	2	Deployment of relational database with one master and one slave
5	Cloud security center	Version: Enterprise Edition	Set	7	Corresponding to the number of servers for system security
6	SSL certificate	Certificate Type: Wildcard DV Version Certificate Brand: Rapid	piece	1	https protocol certificate
7	Domain name	Domain Name	pcs	1	
8	WebRegister	WebRegister	pcs	1	

The software resources required for this project are database, operating system, etc., as shown in the following table.

Serial number	Resource name	Resource type	Resource name	Version	Remarks
1	CentOS	Operating system	Operating system	CentOS 7.4 64 bit	Open source
2	MySQL	Database	Database	mysql-8.0.23	Open source version
3	Redis	Database	Database	redis-3.0.0	Open source version
4	JDK	Runtime environment	Runtime environment	jdk-8u161-linux-x64	Open source version
5	Python	Runtime environment	Runtime environment	Python 3.8.0	Open source version
6	Nginx	Middleware	Nginx	nginx-1.13.9	Open source version
7	Anaconda	Middleware	Anaconda	Anaconda3-2022.05-Linux-x86_64	Open source version

8.4 Initial Investment Estimation

This modeling focuses on calculating the Capital Expenditure (CapEx), which represents the initial investment, for projects involving Distributed Energy Resources (DERs). The goal is to comprehensively account for all the major cost components in setting up Virtual Power Plant (VPP) at the beginning.

For initial investment, three main factors are considered, hardware construction cost, platform construction cost, and operational cost. We use a linear equation to explain the basic relationship between total initial expenditure and the three contributing factors.

Firstly we define the following variables (Table 8.4).

Then, we start modeling the three factors. Hardware construction cost, platform construction cost, and operational cost are defined as follows.

$$\begin{aligned} \text{Hardware construction cost} &= C \times 1000 \times c_{\text{capex}} \\ \text{Platform construction cost} &= c_{\text{platform}} \\ \text{Operational cost} &= N \times c_{\text{device}} \end{aligned} \tag{8.1}$$

c_{capex} represents the capital cost per kilowatt of installed DER capacity, with the unit being USD per kilowatt (USD/kW). It accounts for the cost of hardware, such as the DER equipment itself, and related infrastructure construction per unit of power capacity. C represents the total aggregated capacity of DERs, measured in Megawatts (MW). This indicates the overall size of the DER project in terms

Table 8.4 Capital expenditure (CapEx) variables and definitions

Variable	Symbol	Unit	Definition
Installed capacity	C	MW	Total aggregated distributed energy resources (DERs) capacity under the VPP
Unit capital cost	c_{capex}	USD/kW	Capital cost per kilowatt installed
Platform development cost	c_{platform}	USD	One-time IT platform setup or licensing cost
Device incentive per customer	c_{device}	USD	One-time incentive payment per household/device
Number of customers/devices	N	—	Number of DER participants (households, devices, etc.)

of power-generation potential. Since the installed capacity C is in MW and 1 MW equals 1000 kW, multiplying C by 1000 converts the capacity unit to kilowatts. Then, multiplying by c_{capex}(cost per kilowatt) gives the total capital cost associated with the installed DER capacity. This covers the major hardware and construction costs for the energy-generating part of the project.

c_{platform} is the one-time cost for setting up or licensing the IT platform needed for the DER project, measured in US$s. This could include costs for developing software to manage the DERs, monitor their performance, etc. This is a standalone, one-off cost that is not directly related to the size of the DER capacity or the number of participants. It is a necessary investment for the IT infrastructure to support the DER project.

Operational cost is evaluated by total incentive cost. Incentive cost is defined by multiplying the number of participants by the incentive amount per participant It is used to encourage customer participation, which is crucial for the success of many DER projects, especially those involving distributed residential or small-scale commercial participants.

Finally we calculate the total initial expenditure by multiplying the three contributing factors.

$$\begin{aligned}\text{Total initial expenditure} &= \text{Hardware construction cost}\\ &\quad + \text{Platform construction cost}\\ &\quad + \text{Operational cost}\end{aligned} \tag{8.2}$$

This equation combines the three factors to provide a comprehensive view of the initial investment required for a Virtual Power Plant (VPP) project. It reflects the total capital expenditure needed to set up and operate a DER project, encompassing hardware construction, platform development, and the ongoing operational costs associated with customer participation. This estimate is a critical starting point for project planning, budget allocation, and financial management.

In most cases, the hardware construction is completed, and the platform construction cost is relatively low. The operational cost is determined by the number of customers/devices participating in, as well as the cost of incentives provided to customers.

In summary, the total initial expenditure for a Virtual Power Plant (VPP) project varies with a function of the installed capacity, platform construction cost, and operational cost. This equation provides a solid foundation for estimating the initial capital investment, and it highlights the importance of hardware construction, platform development, and customer participation in the project's success.

8.5 Reduction Revenue Prediction Modeling

8.5.1 Model Construction

In this section, a business model is built for virtual power plants and analyse the benefits.

The overall revenue model aggregates four components, considering real-time data on electricity prices, policy subsidies, load profiles, and carbon market dynamics. By integrating these sub-models, the VPP can predict annual revenues, such as combining demand side response, ancillary services, spot market gains, and environmental/capacity benefits, yielding a total projected revenue. This framework enables data-driven decision-making for optimal VPP operation in one aluminum industrial park.

Firstly we model the revenue from the demand side. This revenue mainly includes two parts, response electricity subsidy and capacity subsidy. Policy-driven subsidies encourage VPPs to shift or reduce industrial loads (e.g., aluminum production) during grid stress, easing pressure on power supply while generating revenue. Storage systems help store excess power (e.g., from renewable sources) and release it during peak hours, reducing reliance on grid power and qualifying for capacity-based incentives.

The following equation shows the modeling of subsidy revenue.

$$R_{\text{subsidy}} = L_{\text{peak}} \times r_{\text{subsidy}} \times N_{\text{response}} + C_{\text{storage}} \times r_{\text{storage}} \tag{8.3}$$

where L_{peak} is the peak load reduction (kW), r_{subsidy} is the subsidy rate per kW per response, and N_{response} is the annual response frequency. C_{storage} is the energy storage capacity of non-independent projects, and r_{storage} is the policy-defined capacity subsidy rate. The first term on the right side of the equation models the revenue of response electricity subsidy, and the second term models the revenue of capacity subsidy.

Then we model the revenue from the ancillary service market. It quantifies earnings from grid support services, divided into three sub-categories, frequency regulation revenue, voltage regulation revenue and reserve service revenue. Aluminum plants' massive loads can fluctuate, so VPPs adjust their consumption in real-time to maintain grid frequency, earning fees for this critical service, and this is where the frequency regulation revenue comes from.

Similarly, VPPs High-power aluminum operations can impact local voltage; VPPs use electricity storage device to stabilize voltage, generating additional income, and we call this part the voltage regulation revenue.

Finally, by keeping spare capacity available, VPPs support grid reliability and are compensated for the readiness, and this part is call the voltage regulation revenue. The following equation shows the modeling of ancillary service revenue.

$$\begin{aligned} R_{\text{freq}} &= M_{\text{reg}} \times r_{\text{freq}} \\ R_{\text{volt}} &= C_{\text{volt}} \times r_{\text{volt}} \\ R_{\text{reserve}} &= C_{\text{res}} \times t_{\text{res}} \times r_{\text{res}} \\ R_{\text{ancillary}} &= R_{\text{freq}} + R_{\text{volt}} + R_{\text{reserve}} \end{aligned} \tag{8.4}$$

where

- M_{reg} = Frequency regulation mileage (MW·Hz)
- r_{freq} = Compensation rate per unit regulation mileage
- C_{volt} = Voltage regulation capacity (MW)
- r_{volt} = Compensation rate per unit voltage capacity
- C_{res} = Reserve capacity (MW)
- t_{res} = Reserve duration (h)
- r_{res} = Compensation rate per MW·h.

Next we model the revenue from the spot market. This revenue mainly includes two parts, electricity sales revenue and purchase cost savings. Aluminum plants can curtail production during high-price periods, storing or selling the saved power to exploit price differences, and the excess energy sold is called the electricity sales revenue. By shifting non-urgent aluminum processes (e.g., preheating) to off-peak hours, VPPs minimize costs, and the money saved is called the purchase cost savings. The following equation shows the modeling of spot market revenue.

$$\begin{aligned} R_{\text{sales}} &= P_{\text{sell}} \times V_{\text{sell}} \\ R_{\text{savings}} &= (P_{\text{peak}} - P_{\text{off-peak}}) \times V_{\text{purchase}} \\ R_{\text{spot}} &= R_{\text{sales}} + R_{\text{savings}} \end{aligned} \tag{8.5}$$

where

- P_{sell} = Spot market selling price ($/kWh)
- V_{sell} = Electricity sales volume (kWh)
- P_{peak} = peak period price
- $P_{\text{off-peak}}$ = Off-peak price
- V_{purchase} = Electricity purchased during low-price periods (kWh).

Finally for the environmental/capacity benefits, we model the environmental benefits from the carbon market. Aluminum smelting is highly carbon-intensive. VPPs can adopt cleaner energy or optimize processes to lower emissions, monetizing credits in carbon markets. Reduced carbon emissions can be sold as carbon credits, the revenue in this part is called the carbon emission trading revenue. Upgrading aluminum production equipment reduces energy use, qualifying for government subsidies tied to carbon intensity targets. Subsidies are awarded for energy savings equivalent to standard coal, and the revenue achieved from this process is called environmental subsidy revenue.

For the capacity benefits, it is made up of two parts, capacity fee expenditure reduction and capacity compensation income. Aluminum plants often have high peak loads. VPPs smooth out demand, reducing the "peak capacity" charged by the grid, thus cutting fixed costs. Industrial users pay a "capacity fee" based on their maximum demand. Reducing peak capacity lowers this fee, and this revenue is the capacity fee expenditure reduction. By intentionally limiting peak demand, VPPs help the grid avoid building extra infrastructure, earning direct compensation for this service. Grid companies compensate VPPs for voluntarily reducing their peak capacity, and this revenue is called the capacity compensation income. The following equation shows the modeling of environmental/capacity benefits.

$$\begin{aligned} R_{\text{carbon}} &= E_{\text{reduction}} \times P_{\text{carbon}} \\ R_{\text{env}} &= S_{\text{coal}} \times r_{\text{env}} \\ R_{\text{cap\_fee}} &= C_{\text{cut}} \times r_{\text{cap\_fee}} \times 12 \\ R_{\text{cap\_comp}} &= C_{\text{cut}} \times r_{\text{cap\_comp}} \end{aligned} \tag{8.6}$$

$$R_{\text{env\_cap}} = R_{\text{carbon}} + R_{\text{env}} + R_{\text{cap\_fee}} + R_{\text{cap\_comp}} \tag{8.7}$$

where

- $E_{\text{reduction}} = CO_2$ emission reduction (tons)
- P_{carbon} = Carbon market trading price
- S_{coal} = Standard coal equivalent (tons)
- r_{env} = Environmental subsidy rate per ton of standard coal
- C_{cut} = Reduced electricity capacity (kVA)
- $r_{\text{cap\_fee}}$ = = Capacity fee rate
- $r_{\text{cap\_comp}}$ = = Capacity compensation rate.

The total revenue equals to the sum of all four categories mentioned above.

$$R_{\text{total}} = R_{\text{subsidy}} + R_{\text{ancillary}} + R_{\text{spot}} + R_{\text{env\_cap}} \quad (8.8)$$

8.5.2 Supporting Example

An industrial park is used as a model for analysis. The production of this industrial park is mainly aluminum, and the consumption of this industrial park is mainly based on electrolytic aluminum, preliminary estimates of the production of each ton of aluminum need to consume 13,000–15,000 kWh of electricity. Others electricity consumption include residential, commercial, and utility power consumption.

Benefits of Virtual Power Plants for Overall Electricity Consumption, Especially in the Aluminum Industry

Demand-side Response Benefits

1. **Subsidy Basis**: Relevant policy documents stipulate that virtual power plants and other market entities participating in demand-side response receive subsidies. These subsidies encourage them to adjust electricity loads, maintaining the balance between electricity supply and demand.
2. **Subsidy Standard**: Calculated as (Peak Load Reduction × USD 1.5/kW/time × Number of Demand-side Response Times in Henan Province per Year). For instance, if a virtual power plant cuts a peak load of 1000 kW during demand-side response, and there are 5 response times in Henan Province in a year, the subsidy amount is $1000 \times 1.5 \times 5 =$ USD 7500.

Capacity Subsidy

New energy projects with non-independent energy storage and user-side non-independent energy storage of 1000 kWh or more are eligible for incentives. For example, a virtual power plant in an industrial park can earn USD 60,000 per year (converted from RMB 400,000 at an assumed exchange rate) through demand response.

Ancillary Service Market Gain

1. **Frequency Modulation (FM) Service Revenue**: Calculated based on FM mileage or FM capacity. If a virtual power plant has an FM mileage of 80 MW.Hz and the unit FM mileage compensation price is USD 6/MW.Hz (converted from RMB 40), the FM service revenue is $80 \times 6 =$ USD 480. For FM capacity compensation, it is calculated by multiplying the provided FM capacity by the unit FM capacity compensation price.
2. **Voltage Regulator Service Revenue**: Determined by factors like voltage regulation capacity or duration. If the unit FM capacity compensation price is USD 5.2/MW (converted from RMB 35) and a virtual power plant provides 60 MW of FM capacity, the revenue is $5.2 \times 60 =$ USD 312.

3. **Standby Service Revenue**: Calculated according to standby capacity and time. Assuming the compensation price per unit of standby capacity and time is USD 3.75/MW/hour (converted from RMB 25), and a virtual power plant provides 15 MW of standby capacity for 6 h, the standby service revenue is $3.75 \times 15 \times 6 =$ USD 337.5.

Electricity Spot Market Trading Revenue

Henan industrial and commercial time-sharing tariffs for peak and valley have a floating ratio: peak, flat-section, and low-valley ratios are adjusted to 1.72:1:0.45; the peak floating ratio is 1.2 times that of the flat section, meaning the peak electricity price is 2.064 times the flat-section price, creating a significant price difference.

1. **Revenue from Electricity Sales**: Virtual power plants sell electricity in the spot market to the grid or other players. Revenue equals the selling price per kWh multiplied by the electricity sold. For example, selling 5000 kWh at USD 0.075/kWh (converted from RMB 0.5) to the grid generates revenue of $0.075 \times 5000 =$ USD 375.
2. **Purchase Cost Savings**: Through optimal scheduling, virtual power plants buy more power at low prices in the spot market, reducing high-price grid purchases. Savings equal the cost difference. If high - price electricity is USD 0.12/kWh (converted from RMB 0.8) and low-price is USD 0.045/kWh (converted from RMB 0.3), and 1000 kWh is bought at the low price, the savings are $(0.12 - 0.045) \times 1000 =$ USD 75.

Energy Saving and Emission Reduction Gains

As the carbon emissions trading market develops, emissions reduced by enterprises via virtual power plants become economic gains.

1. **Carbon Emission Right Trading Income**: First, calculate reduced carbon emissions, then multiply by the carbon emission right market price. If a virtual power plant reduces 800 tons of carbon dioxide emissions (through optimized energy use) and the carbon price is USD 6/ton (converted from RMB 40), the revenue is $800 \times 6 =$ USD 4800.
2. **Environmental Protection Subsidy Income**: Based on government policies. If a virtual power plant meeting energy-saving and emission-reduction standards gets a subsidy of USD 12/ton of standard coal (converted from RMB 80) and achieves an effect equivalent to 300 tons of standard coal, the subsidy is $300 \times 12 =$ USD 3600.

Capacity Compensation Gain

1. **Reduction of Capacity Electricity Expenditure**: Calculate the reduced electricity capacity after virtual power plant load management optimization, then multiply by the unit capacity electricity price. (Example continuation can be added based on complete original content if needed.)

8.6 Quantitative Risk Analysis

Virtual power plants face market risks (electricity spot price fluctuation and carbon price volatility), policy and regulatory risks (reduced demand response and capacity subsidies), and technical and operational risks (load response failure and energy storage efficiency decline).

8.6.1 *Market Risk Analysis*

(1) Electricity Spot Price Fluctuation

The revenue from the electricity spot market is highly sensitive to price volatility. Given the peak-valley price difference model (e.g., peak price = 2.064 × flat price), sudden price swings can impact both selling profits and purchasing cost savings.

$$\Delta R_{\text{spot}} = Q \times (\Delta P_{\text{peak}} - \Delta P_{\text{valley}}) \tag{8.9}$$

where ΔR_{spot} is the revenue variation, Q is the electricity volume, and $\Delta P_{\text{peak}} / \Delta P_{\text{valley}}$ are price fluctuations in peak/valley periods. If the peak price drops by 10% and the valley price rises by 5%, a virtual power plant (VPP) with 10,000 kWh trading volume would face a revenue loss of:

$$10{,}000 \times (0.1 \times 2.064 P_{\text{flat}} + 0.05 \times 0.45 P_{\text{flat}}) = 2397 P_{\text{flat}}$$

Carbon Price Volatility

Carbon trading revenue depends on carbon price dynamics. For instance, if the carbon price drops from 6 USD/ton to 4.5 USD/ton (converted from 40 CNY/ton to 30 CNY/ton at an approximate rate of 1 CNY = 0.15 USD), the revenue from 800 tons of emission reductions would decrease by:

$$800 \times (6 - 4.5) = 1200 \text{ USD}$$

8.6.2 *Policy and Regulatory Risk*

- Demand response subsidies: Reductions in peak load subsidy rates directly impact revenue:

$$\Delta R_{\text{DR}} = -1000 \times (1.5 - 0.75) \times 5 = -3750 \text{ USD/year}$$

- Capacity subsidies: Cuts in incentives for energy storage projects (e.g., scaling back rewards for projects >1000 kWh) could eliminate 60,000 USD/year in expected revenue.

8.6.3 Technical and Operational Risk

(1) Load Response Failure

If the VPP fails to adjust 1000 kW of peak load due to technical issues, the demand response subsidy loss is:

$$1000 \times 1.5 \times 5 = 7500 \text{ USD/year} \quad \text{(converted from 10 CNY/kW} \cdot \text{time to 1.5 USD/kW} \cdot \text{time)}$$

Energy Storage Efficiency Decline

A 10% drop in energy storage efficiency (e.g., from 90 to 80%) would reduce spot market arbitrage profits. Assuming peak-valley price differences in USD (converted from CNY), the reduction is:

$$\Delta R_{\text{arbitrage}} = 1000 \times (0.12 - 0.045) \times 0.1 = 7.5 \text{ USD/transaction}$$

(Note: 0.8 CNY/kWh = 0.12 USD/kWh and 0.3 CNY/kWh = 0.045 USD/kWh at 1 CNY = 0.15 USD)

8.6.4 Risk Mitigation Strategies

- Diversification: Expand into frequency regulation (e.g., 80 MW·Hz × 6 USD = 480 USD per service) and voltage regulation markets to offset spot price risks.
- Hedging: Use financial instruments (e.g., electricity futures) to lock in 60–70% of spot market volumes.
- Policy Tracking: Allocate 3–5% of revenue to monitor regulatory changes, ensuring timely adaptation to subsidy adjustments.
- Technical Redundancy: Maintain 15–20% excess energy storage capacity to mitigate response failures.

8.7 Significance of Virtual Power Plant Construction with Financial Benefit Analysis

The construction of virtual power plants (VPPs) holds multi-faceted significance, integrating operational efficiency, grid stability, and financial sustainability. By aggregating distributed energy resources (DERs) such as on-site generation, energy storage, and flexible loads, VPPs enable optimized energy management and create new revenue streams for industries like aluminum processing [1–3].

1. Optimizing Energy Utilization

VPPs enhance enterprise energy management by integrating waste heat recovery, on-site power plants, and energy storage systems. For aluminum enterprises, this integration enables intelligent scheduling to match energy supply with demand, reducing operational costs. For example, waste heat from smelting processes can be repurposed for preheating raw materials, lowering fossil fuel consumption and associated expenses. Financial benefits include reduced energy procurement costs and potential tax incentives for energy-efficient upgrades.

2. Reducing Electricity Costs

Participation in power market transactions allows VPPs to capitalize on price differentials. Aluminum plants can store energy during low-price periods and either use it during peak hours or sell it back to the grid, generating arbitrage revenue. For instance, with peak-valley price ratios of 2.064:1, strategic energy trading can yield significant savings. Additionally, backup power from VPPs mitigates losses from grid outages, which can cost industrial facilities millions in production downtime and equipment damage.

3. Enhancing Grid Stability

By providing grid services such as frequency regulation and peak shaving, VPPs qualify for ancillary service revenues. For example, a VPP providing 80 MW·Hz of frequency regulation at USD 6/MW·Hz earns USD 480 per service. Aluminum plants with large, stable loads can monetize their flexibility by participating in demand response programs, receiving subsidies for load curtailment (e.g., USD 1.5/kW·time in subsidy for reducing peak loads).

4. Promoting Energy Conservation and Emission Reduction

VPPs facilitate the integration of renewable energy sources, reducing carbon emissions and enabling participation in carbon markets. For example, reducing 800 tons of CO_2 emissions at a carbon price of USD 6/ton generates USD 4800 in trading revenue. Additionally, energy efficiency improvements may qualify for government subsidies (e.g., USD 12/ton of standard coal saved), further enhancing financial returns.

5. Driving Industrial Upgrading

VPPs enable aluminum enterprises to diversify revenue streams by offering energy management services to neighboring industries. For example, providing energy hosting or consulting services can create new income sources. Collaborations with grid operators and technology providers also accelerate innovation, enhancing the competitiveness of aluminum manufacturing. VPPs also reduce long-term operational costs through shared R&D investments and improved energy infrastructure.

References

1. Zeng, M., et al. (2016). Coordinated optimization operation mode and key technologies of "source-grid-load-storage" in energy internet. *Power System Technology, 40*(01), 114–124.
2. Hadayeghparast, S., SoltaniNejad Farsangi, A., & Shayanfar, H. (2019). Day-ahead stochastic multi-objective economic/emission operational scheduling of a large scale virtual power plant. *Energy, 172*, 630–646.
3. Zhang, J., Nong, J., Xu, Z., et al. (2018). A costs-emissions bi-objective optimization of virtual power plant operation in Hongfeng eco-town. In *2018 37th Chinese control conference (CCC)* (pp. 8752–8757).

Part III
Go Smarter

Get to know more about *Virtual Power Plant* on AI and future trends.

Chapter 9
AI Within Reach

Abstract Virtual Power Plants (VPPs) are cloud-based networks that aggregate distributed energy resources (DERs)—such as solar panels, wind turbines, and battery storage—into a unified system, enabling flexible, decentralized energy management superior to traditional centralized power plants. By integrating artificial intelligence (AI) and Large Language Models (LLMs), VPPs achieve enhanced efficiency, resilience, and adaptability, becoming critical to modern energy systems. AI drives VPP operations through three key functions: predictive analytics, which forecasts energy demand, renewable generation, and market prices using multi-source data; optimization algorithms (e.g., reinforcement learning) that dynamically schedule DERs to balance supply and demand, minimizing costs; and anomaly detection, which monitors grid health in real time to prevent disruptions. LLMs complement AI by facilitating human-AI interaction via natural language processing (NLP), enabling intuitive control for operators; automating reporting on energy trends, DER performance, and regulatory compliance; and providing decision support to navigate complex policies. Together, these technologies form the "core intelligence" of VPPs, integrating forecasting, optimization, communication, and risk management. Future advancements will focus on autonomy (via multi-agent reinforcement learning), resilience (through generative AI scenario simulations), and privacy (using federated learning). These developments will transform VPPs into self-learning, adaptive networks, solidifying their role in integrating renewables, stabilizing grids, and advancing sustainable energy transitions.

9.1 Go with AI

9.1.1 Virtual Power Plants (VPPs) and the Role of AI/LLMs

A **Virtual Power Plant (VPP)** is a sophisticated cloud-based network that aggregates and coordinates distributed energy resources (DERs)—including solar panels, wind turbines, battery storage systems, and demand-response-capable devices—into a unified, intelligent system. Unlike traditional power plants, which rely on centralized generation, VPPs leverage the flexibility and scalability of decentralized

F. Yao, *Engineering Guide to Planning and Design of Virtual Power Plant*,
https://doi.org/10.1007/978-981-95-2231-6_9

resources to optimize energy flow, enhance grid resilience, and adapt to dynamic market conditions [1].

By integrating advanced technologies such as artificial intelligence (AI) and Large Language Models (LLMs), VPPs can achieve unprecedented levels of efficiency, responsiveness, and operational intelligence, making them a cornerstone of modern energy systems.

AI-Driven Enhancements in VPP Operations

Artificial intelligence plays a pivotal role in transforming VPPs from passive resource aggregators to proactive energy managers, with three key applications driving operational excellence. First, **predictive analytics** empowers VPPs to forecast critical variables with high accuracy: AI models analyze historical data, weather patterns, and market trends to predict energy demand, the output of renewable sources like solar and wind, and fluctuations in electricity prices. These forecasts enable VPP operators to anticipate supply-demand gaps, adjust resource allocation in advance, and capitalize on favorable market conditions.

Second, **optimization algorithms** dynamically schedule the deployment of DERs to maximize efficiency and minimize costs. For example, during periods of high renewable generation (e.g., sunny or windy days), AI can prioritize using solar or wind energy, store excess power in batteries, or redirect it to meet immediate demand—reducing reliance on expensive grid electricity or fossil fuels. Conversely, during peak demand, the algorithm can activate demand-response systems, incentivizing users to reduce consumption temporarily, thus balancing the grid and avoiding costly peak-time charges.

Third, **anomaly detection** allows AI to monitor DER performance and grid stability in real time. By identifying deviations from normal operating patterns—such as a sudden drop in solar panel output, a battery malfunction, or signs of grid instability—AI triggers alerts or automated responses, minimizing downtime, preventing cascading failures, and ensuring the reliability of the entire VPP network.

Large Language Models (LLMs) in VPPs

Large Language Models (LLMs) extend the capabilities of VPPs by bridging the gap between complex technical systems and human operators, while also streamlining administrative and decision-making processes. A primary application is **Natural Language Processing (NLP)**, which enables intuitive human–AI interaction. Grid operators, for instance, can query energy forecasts, check the status of specific DERs, or adjust operational parameters using conversational interfaces or chatbots—eliminating the need for specialized technical training and speeding up response times.

LLMs also excel at **automated reporting**, generating clear, concise summaries of critical metrics such as energy market trends, DER performance metrics (e.g., solar panel efficiency or battery discharge rates), and compliance with regulatory requirements. This automation reduces the administrative burden on operators, allowing them to focus on strategic tasks rather than manual data compilation.

Additionally, LLMs provide valuable **decision support** by interpreting complex energy policies, market regulations, and incentive programs. For example, when navigating changes in demand-response subsidy schemes or carbon pricing policies, LLMs can analyze the implications for the VPP, suggest optimal strategies to maximize revenue or minimize costs, and even draft proposals for compliance—empowering operators to make informed choices in a rapidly evolving regulatory landscape.

Together, AI and LLMs transform VPPs into adaptive, intelligent systems that not only manage energy resources efficiently but also integrate seamlessly with human operations and regulatory frameworks, paving the way for a more resilient, sustainable, and cost-effective energy future.

9.1.2 Core Intelligence Components

Artificial Intelligence (AI) and Large Language Models (LLMs) function as the "brain" of a Virtual Power Plant (VPP), seamlessly integrating into its key operational components to enhance efficiency and decision-making.

Forecasting, a critical pillar, relies on AI to predict energy demand, renewable generation (such as solar and wind output), and electricity market prices with high accuracy, enabling proactive resource allocation. Optimization is driven by Reinforcement Learning (RL) algorithms, which dynamically schedule Distributed Energy Resources (DERs)—including batteries, solar panels, and demand-response systems—to maximize energy utilization, minimize costs, and balance supply and demand.

Anomaly detection, another vital component, leverages AI to continuously monitor grid health, identifying potential issues such as equipment malfunctions, voltage fluctuations, or unusual energy consumption patterns in real time, thus preventing disruptions. Communication is streamlined through LLMs, which facilitate intuitive human-AI interaction: grid operators can troubleshoot issues, query system statuses, or communicate policy updates via natural language interfaces, reducing operational complexity.

Finally, reporting and decision support are enhanced by LLMs, which generate real-time insights into energy market trends, DER performance, and regulatory compliance, equipping operators with actionable information to make strategic choices.

9.1.3 Future Improvements

AI and LLMs are poised to evolve further, making VPPs increasingly autonomous and efficient, with advancements focused on three key areas.

Autonomous Grid Management will see AI agents self-optimize DERs without human intervention, powered by Multi-Agent Reinforcement Learning (MARL). In this framework, individual agents representing different DERs or grid segments will collaborate, learn from each other, and adapt to dynamic conditions—such as sudden demand spikes or renewable generation drops—ensuring seamless, real-time balancing of the grid.

Generative AI, particularly advanced LLMs like GPT-4, will enable Scenario Simulation, allowing VPPs to model and prepare for extreme grid conditions. These models can simulate events such as cyberattacks, severe storms, or prolonged supply shortages, helping operators design contingency plans, strengthen infrastructure, and improve overall system resilience.

Federated Learning will address privacy concerns by enabling AI models to train on decentralized DER data—such as household energy consumption or industrial load patterns—without exposing sensitive consumer or operational information. This approach ensures that AI systems improve their accuracy over time while safeguarding data privacy, a critical consideration in increasingly interconnected energy networks.

9.1.4 Future Advancements

AI and LLMs are transforming VPPs into self-learning, adaptive energy networks, with future advancements focusing on enhancing autonomy, resilience, and explainability. As these technologies mature, VPPs will become more capable of operating independently, responding to disruptions, and clearly communicating their decision-making processes to operators and regulators. This evolution will solidify VPPs as a cornerstone of the smart grid, playing a pivotal role in integrating renewable energy, ensuring grid stability, and supporting sustainable energy transitions.

Reference

1. China Huadian Engineering Group Co., Ltd. (Ed.). (2025). *Clean energy and green computing synergy*. China Electric Power Press.

Appendix A
Formula, Diagram and Function Block

A.1 Types of Distributed Energy Resources

The types of resources aggregated by virtual power plants mainly include the following:

A.1 Distributed Energy Resources
Natural gas distributed power generation units;
Distributed photovoltaic power generation units;
Distributed wind power generation units;
Small hydroelectric generating units.
A.2 Energy storage devices
New energy electric vehicles and charging piles (two-way charging and discharging);
Centralized energy storage devices;
Decentralized energy storage devices.
A.3 Controllable load
New energy electric vehicle charging piles (one-way charging);
Ice and heat storage devices;
Loads installed with smart actuators.

1. Distributed generation resources

   (1) Solar photovoltaic power generation: including distributed rooftop photovoltaic power generation systems and ground-mounted photovoltaic power stations. These photovoltaic power generation facilities vary in size. Small ones can be roof photovoltaic panels of residential houses or small commercial buildings, while large ones can be centralized photovoltaic power stations in industrial parks. For example, photovoltaic panels installed on the roofs of plants in some industrial parks convert solar energy into electricity during the day, and their power generation is affected by factors such as light intensity and weather conditions.

F. Yao, *Engineering Guide to Planning and Design of Virtual Power Plant*,
https://doi.org/10.1007/978-981-95-2231-6

(2) Wind power generation: there are decentralized small wind turbines and centralized wind farms. Small wind turbines are suitable for areas with relatively stable wind speed and small power demand, such as remote farms or small facilities by the sea; large-scale wind farms are composed of multiple wind turbines, whose power generation is directly proportional to the cube of the wind speed, and are generally built in open areas such as coastal areas or plateaus where wind resources are abundant.
(3) Biomass power generation: the use of biomass (such as crop residues, forestry waste, organic waste, etc.) through combustion or biochemical conversion to generate electricity. For example, a biomass power plant fueled by crop residues first pre-treats the residues and then burns them to produce steam to drive a turbine to generate electricity. Its power generation capacity depends on factors such as the stability and quality of the biomass supply.
(4) Small hydropower generation: usually refers to hydropower plants with smaller installed capacity, such as small hydropower in rural areas. These small hydropower utilizes the water energy resources of rivers, streams, etc., and converts water energy into electricity through hydraulic turbines. Their power generation capacity is related to the flow and fall of the water current, and can be affected by seasons and precipitation.

2. Energy storage resources

(1) Electrochemical energy storage: lithium-ion battery energy storage: with the advantages of high energy density, low self-discharge rate and long cycle life, it is widely used in energy storage systems of various scales. From home energy storage systems to large-scale grid-level energy storage power plants are involved. For example, in some distributed photovoltaic power generation systems, the supporting lithium-ion battery energy storage system can store excess photovoltaic power generation during the day and release the power at night or when the light is insufficient.
(2) Lead-acid battery energy storage: mature technology, relatively low cost, commonly used in some of the energy density requirements are not high, more sensitive to the cost of energy storage scenarios, such as small UPS (uninterruptible power supply) system or small energy storage systems in remote areas. Fluid flow battery energy storage: the energy storage and release process is realized through the redox reaction of active substances in the electrolyte, and its advantage is that the capacity can be flexibly adjusted by increasing the volume of electrolyte. Suitable for long time energy storage or large capacity energy storage occasions, such as large-scale grid energy storage or renewable energy power station energy storage package.
(3) Pumped storage: a more mature large-scale energy storage technology. By pumping water from the lower reservoir to the upper reservoir to store it when the power load is low, the water in the upper reservoir is released to drive the turbine to generate electricity when the power load is high. Its energy storage capacity is large, but it is limited by geographical conditions and requires suitable topography of the upper and lower reservoirs. Compressed air energy

storage: In the power trough, the excess power is utilized to compress and store air in underground caves or high-pressure containers; in the power peak, the compressed air is released to drive gas turbines to generate electricity. The efficiency of this type of energy storage is related to factors such as the sealing of the storage container and the pressure of the compressed air.

3. Controllable load resources

(1) Industrial loads: The electricity loads of large motors, furnaces and other equipment in industrial enterprises can be adjusted according to the demand of the power grid. For example, electrolytic aluminum enterprises can appropriately reduce the working current of electrolysis tanks and reduce the power consumption when the load of the grid is at its peak; iron and steel enterprises can adjust the production rhythm of steel-making furnaces to respond to the scheduling instructions of the grid without affecting the production plan. Commercial Load: The air conditioning system and lighting system of large shopping malls and office buildings are loads that can be adjusted. Through the intelligent control system, the air-conditioning temperature, lighting brightness, etc. can be adjusted according to the high or low price of electricity or the demand of the power grid without affecting the indoor comfort, so as to realize the flexible control of the load.
(2) Residential load: It mainly includes electrical appliances controlled by smart home devices, such as washing machines and water heaters controlled by smart meters and smart sockets. Through incentives such as time-sharing tariffs, residents are guided to use these appliances during the trough tariff hours, or the power consumption of these appliances is reduced within a short period of time according to the emergency demand of the power grid, so as to realize the aggregated regulation of residential loads. Electric Vehicle Resources
(3) Electric vehicle charging facilities: including private charging piles, public charging piles and power exchange stations. When a large number of electric vehicles are connected to the charging facilities, these facilities can be used as regulation resources for virtual power plants. For example, through the intelligent charging management system, it can speed up the charging speed of electric vehicles when the grid load is low, and reduce the charging power or suspend charging when the grid load is peak, so as to realize the regulation of the grid load.
(4) Energy storage potential of electric vehicle batteries: The batteries of electric vehicles themselves also have energy storage functions. In the future, with the development of Vehicle-to-Grid Interaction (V2G) technology, electric vehicles can feed back the power in the battery to the power grid when parked, as a mobile energy storage unit, providing frequency and voltage regulation and other ancillary services for the power grid.

A.2 Comprehensive Benefit Optimization Evaluation Formula Related to Virtual Power Plant

The comprehensive benefit optimization evaluation shall satisfy the following objective function to maximize the virtual power plant power generation benefit and satisfy the constraints.

A.1.1 Objective Function

$$\max P_{profit} = = P^{out}_{profit} + P^{in}_{profit} - C^{market}_{punish} - C_{dq} - C_{wind} - C_{PV} - C_{bat} - C_{DR} \tag{A.1.1}$$

$\max P_{profit}$ is the revenue from power generation, and

P^{out}_{profit} is the virtual power plant's gain from selling electricity in the electricity market, and

P^{in}_{profit} is the virtual power plant's revenue from the virtual power plant's internal power sales, and

C^{market}_{punish} is the compensation if the actual output does not meet the contractual requirements.

C_{dq} is the total cost of the gas turbine, including operating costs, startup and shutdown costs, and standby costs.

C_{wind} is the sum of wind abandonment costs and operation and maintenance costs for wind power.

C_{PV} is the sum of the abandoned light cost and operation and maintenance cost of photovoltaic power generation.

C_{bat} is the cost of charging and discharging energy storage devices.

C_{DR} is the demand side response cost of the virtual power plant.

A.1.2 Constraints to be met by the virtual power plant

A.1.2.1 Gas turbine constraints

$$\begin{aligned} P^{min}_{d,g,i} &\leqslant P^{t}_{d,g,i} + R^{t}_{d,g,i} \leqslant P^{max}_{d,g,i} \\ P^{down}_{d,g,i} &\leqslant P^{t+1}_{d,g,i} - P^{t}_{d,g,i} \\ &\leqslant P^{up}_{d,g,i} \end{aligned}$$

$$(A.1.2.1)$$

$P_{d,g,i}^{min}$ and $P_{d,g,i}^{max}$ are the lower and upper limits of the power of the ith gas turbine, respectively.

$P_{d,g,i}^{down}$ and $P_{d,g,i}^{up}$ are the downward climb power limit and upward climb power limit for the ith gas turbine, respectively.

$P_{d,g,i}^{t}$ is the power of the ith gas turbine at moment t, and

$R_{d,g,i}^{t}$ is the standby capacity of the ith gas turbine at time t.

A.1.2.2 Wind and Photovoltaic Power Generation Constraints

$$\begin{aligned} 0 \leqslant P_{wind,i}^{t} \leqslant P_{wind,i}^{t,pre} \\ 0 \leqslant P_{provided,i}^{t+1} \leqslant P_{PV,i}^{t,pre} \end{aligned} \quad (A.1.2.2)$$

$P_{wind,i}^{t}$ and $P_{wind,i}^{t,pre}$ respectively the adjustable power of the ith wind turbine at moment t and the predicted adjustable power of the ith wind turbine at moment t.

$P_{provided,i}^{t+1}$ and $P_{PV,i}^{t,pre}$ respectively the adjustable power and the predicted regulated power of the ith PV generating unit at moment t.

A.1.2.3 Energy storage charging and discharging constraints

$$\begin{aligned} SOC_{i}^{min} \leqslant SOC_{bat,i}^{t} \leqslant SOC_{i}^{max} \\ SOC_{bat,i}^{t} = SOC_{bat,i}^{t-1} + P_{bat,i}^{t} \\ P_{bat,i}^{discharge} \leqslant P_{bat,i}^{t} \leqslant P_{bat,i}^{charge} \end{aligned} \quad (A.1.2.3)$$

SOC_{i}^{min} and $\leqslant SOC_{i}^{max}$ are the minimum and maximum energy storage capacity of the ith energy storage device, respectively.

$SOC_{bat,i}^{t}$ is the energy storage capacity of the ith energy storage device at time t, and

$P_{bat,i}^{discharge}$ and $P_{bat,i}^{charge}$ respectively are the maximum discharge power and the maximum charge power of the ith energy storage device, and

$P_{bat,i}^{t}$ is the discharge power of the ith energy storage device at moment t.

A.1.2.4 Demand-side response constraints

$$0 \leqslant P^t_{curt,i} \leqslant P^{t,max}_{curt,i}$$

$$\sum_{i=1,\in VPP}^{n} -P^t_{curt,i} = P^t_{curt}$$

$$(A.1.2.4)$$

$P^t_{curt,i}$ is the interruptible load power of the ith demand-side response user at time t, and

P^t_{curt} is the interruptible load power of all demand-side response users at time t, and

$P^{t,max}_{curt,i}$ is the upper limit of the interruptible load power of the ith demand-side response user at time t, and

N is the number of demand-side response users.

A.1.2.5 System power balance constraints

$$P^t_{dg} + P^t_{wind} + P^t_{PV} + P^t_{bat} + P^t_{curt} = P^t_{load} + P^t_m \quad (A.1.2.5)$$

$P^t_{dg}$is the generation power of the gas turbine at time t

P^t_{wind} is the generation power of the wind turbine at time t, and

P^t_{PV} is the generation power of the photovoltaic generator at time t, and

P^t_{bat} is the discharge power of the energy storage device at time t.

P^t_{curt} is the interruptible load power of the demand-side response user at time t, and

P^t_{load} and P^t_m are the internal load power of the virtual power plant and the power provided by the virtual power plant to the power trading market, with positive values representing the output power to the power trading market and negative values representing the input power from the power trading market.

A.1.2.6 Line current and bus voltage constraints

$$P^t_{ij} \leqslant P^{max}_{ij}$$

$$U^{min}_i U^t_i \leqslant U^{max}_i$$

$$(A.1.2.6)$$

P_{ij}^{t} is the transmission power of the branch circuit between nodes i, j, and

U_i^{max} is the transmission power limit of the branch between nodes i, j, the

U_i^{min} and U_i^{max} are the lower and upper limits of the node voltage, respectively, and

U_i^t is the node voltage.

A.3 Layered Composition of the Virtual Power Plant Design

Virtual power plant should be divided into four layers, layer of distributed energy resources, layer of communication network, layer of data hardware platform for database, layer of intelligent applications (Fig. A.1).

The diagram illustrates a four-layer smart energy system architecture, showing the integration of energy flow, communication networks, data processing, and intelligent applications.

Layered composition of the virtual power plant design

Layer 1: Distributed Energy Resources Layer (Bottom)

Physical infrastructure: Smart grids, digital agriculture power stations, battery storage, and flexible distribution networks. Key tech: Edge control, smart sensors, and renewable energy integration.

Layer 2: Communication Network Layer

Technologies: 4G/5G, NB-IoT, fiber optics, and IoT for real-time data transmission. Role: Connects devices and enables data flow between layers.

Layer 3: Data Hardware Platform Layer

Function: Data collection, cleaning, classification, modeling, and storage (using AI, blockchain, and cloud). Purpose: Transforms raw data into actionable insights.

Layer 4: Intelligent Application Layer (Top)

Smart services: Distributed energy management, power market optimization, smart inspections, and consulting. goal: Supports decision-making and automation in energy operations.

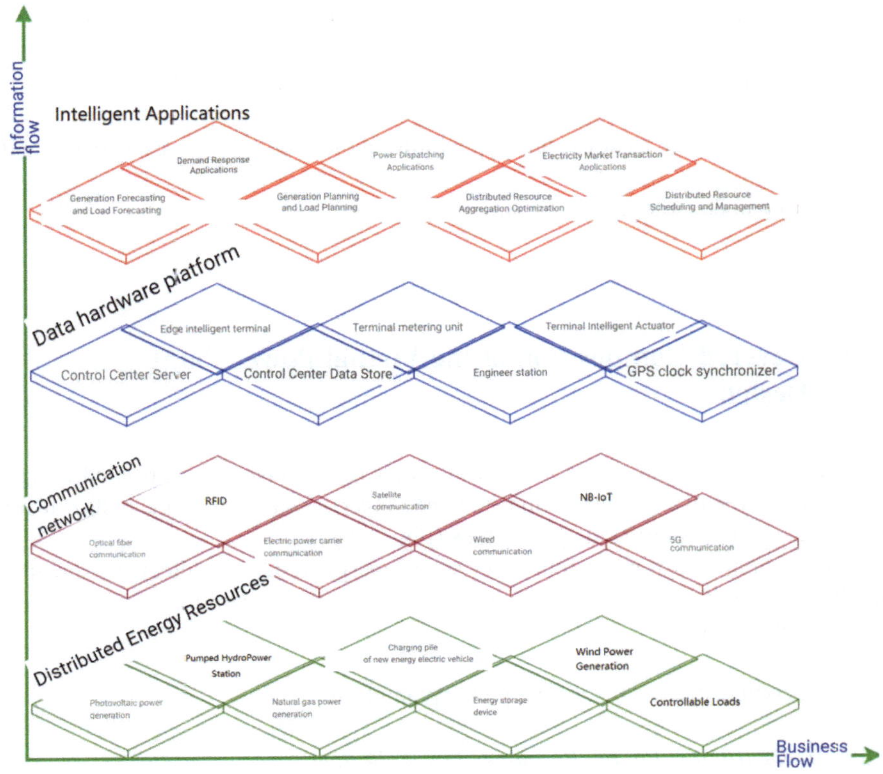

Fig. A.1 Layered architecture of the virtual power plant

Key Flows

Energy Flow: Bottom-up (physical to digital). Information/Business Flow: Top-down (applications guide infrastructure via data). Visual Style: Red/white color contrast with clear labels, emphasizing layer interactions for a modern, AI-driven energy ecosystem.

This architecture enables a fully responsive, decentralized, and data-powered energy grid.

A.4 Design Requirements for Intelligent Function Modules

The design of the intelligent function module of the virtual power plant should meet the following requirements:

A.4.1 Generation Forecast and Load Forecasting

The power generation forecast of the virtual power plant shall predict the power generation of wind power and photovoltaic power generation based on short-term and long-term meteorological data and meteorological forecast information, as well as historical operation data.

The load forecast of the virtual power plant should analyze the industrial production, residential life, weather changes and other factors, as well as the historical operation data, to predict the change of electric load.

The power generation forecast and load forecast of the virtual power plant should be based on high-precision numerical weather forecast and other data, as well as historical operation data, relying on a rich model library of machine learning algorithms, and should realize power generation forecasts in various time dimensions, such as medium- to long-term, medium- to short-term, and ultra-short-term.

For the power generation forecast of virtual power plant, the load forecast needs to meet the requirement of accuracy.

Load forecasting for virtual power plants should be based on many factors such as system operating characteristics, capacity increase decisions, natural conditions and social impacts, etc., and future values should be projected based on the past and present of the power load.

A.4.2 Preparation of Generation Plan and Consumption Plan

The preparation of the generation plan and electricity consumption plan of the virtual power plant shall be planned on the basis of optimization and development of the generation plan and electricity consumption plan of the distributed energy resources belonging to the virtual power plant.

The power generation plan of the virtual power plant shall be based on the load forecast, and under the premise of meeting the power balance, combined with the upper and lower limits of the output of each distributed power generation resource, the maximum ramping power of each unit and other constraints, and taking into account the start and stop of each distributed power generation resource as well as the minimum shutdown time and other actual conditions, the operation plan of power generation in chronological order shall be compiled in advance.

The energy storage device of the virtual power plant can be discharged to the grid when it is fully charged and shall be scheduled in the power generation plan.

The power generation plan of the virtual power plant shall be compiled hierarchically in accordance with the day-ahead power generation plan (one or more days in the future), the intra-day power generation plan (1 to several hours in the future) and the real-time power generation plan (5 or 15 min to 1 h in the future).

The load power plan of the virtual power plant is developed by managing the controllable loads and developing the load plan in a hierarchical manner based on the importance of the controllable loads, load type, availability, and adjustable range. Energy storage devices operating in the charging state, including centralized energy storage devices and decentralized new energy electric vehicles connected to charging piles, etc., are controllable loads and shall be operated in accordance with the load plan.

The load plan of the proposed power plant shall be prepared hierarchically according to day-ahead, intra-day and real-time load variations.

A.4.3 Controllable Load Management

The controllable load management of the virtual power plant shall be based on the adjustable capacity of all controllable loads to meet the requirements of grid operation scheduling, and the part of controllable loads shall be put into or removed in time.

The controllable load management of the virtual power plant should be realized in a market-oriented way with demand-side management and demand-side response management. The virtual power plant should be connected to a large number of adjustable capacity, decentralized, small-scale industrial power loads, non-industrial residential power loads, as controllable loads.

The management of controllable loads should meet the peak and valley tariff difference mechanism, according to the specific peak and valley tariff difference, dynamic allocation of controllable loads to obtain the lowest comprehensive cost of electricity.

The management of controllable loads shall be able to demand response, and the portion of controllable loads connected to the virtual power plant that has the characteristics of rapid interruption or remote interruption shall participate in demand response in accordance with the market rules and processes of demand response, control the participation of the belonging controllable loads in the execution of demand response, and settle the compensation costs based on the evaluation of the response effect.

A.4.4 Control and Management of Energy Storage Devices

The virtual power plant control and management of energy storage devices shall continuously realize the control of energy conversion and power generation capacity, and according to the characteristics of different energy storage devices, shall meet the different requirements of independent energy storage, grid-side energy storage and user-side energy storage.

The virtual power plant control and management of energy storage devices, mainly user-side energy storage devices, meets the technical conditions of having independent metering and control, and can be connected to the dispatch automation system.

Its control and management, should be mainly in accordance with the user side of the energy storage mode of operation, to obtain the difference between peak and valley tariffs as revenue.

In the market mechanism with demand response, it meets the requirement of part of the capacity of the energy storage device as the regulation of the controllable load, with the power generation state and the power consumption state, and the participation in ancillary services such as demand-side response.

The energy storage device of the virtual power plant shall be operated according to the time-sharing plan of the charging and discharging cycle, and after deducting the conversion losses, it shall obtain the benefit of the peak and valley point price difference.

The energy storage device of the virtual power plant shall provide ancillary services such as active balancing service, reactive power balancing service and accident emergency response and restoration service, as well as fast active response service in the event of a grid accident, in accordance with the rules of the ancillary service market or the rules governing ancillary services, and obtain the fees for ancillary services.

A.4.5 Coordination and Optimization of Distributed Generation

Coordination and optimization of distributed generation shall be satisfied to enable the distributed power sources, energy storage devices and controllable loads connected to the virtual power plant to be integrated for real-time coordination and optimization of distributed generation.

The coordination and optimization function should combine all real-time data of distributed power sources, energy storage devices and controllable loads, combined with the results of power generation forecasts and load forecasts, and follow the rules of commercial agreements as well as market trading inquiries or bidding to optimize and obtain the optimal benefits.

The coordinated optimization of distributed generation can be done in a central control center, or a central data exchange and data processing center.

The optimal scheduling strategy for distributed generation should be based on the characteristics and capacity of the resources connected to the virtual power plant, and different weights should be assigned to them in different levels, so as to realize the optimal selection under the constraints according to the sequential order.

The coordinated optimization of distributed generation should satisfy the integrated and cooperative scheduling of wind power, photovoltaic and energy storage systems, and smooth out the system fluctuations introduced by renewable energy sources.

Coordinated optimization of distributed power generation should meet the integrated energy system of combined cooling, heating and power supply, the complementary problems of cooling, heating and power between different regions, expand the single-region heat and power coordinated scheduling optimization problem to multi-region, consider the flexibility problems caused by the strategy of combined heat and power and heat to set power, and increase the power-to-gas optimization unit by taking into account the factors of the natural gas market with a large variation, and so on.

The coordinated optimization of distributed power generation should meet the problems of disordered charging and discharging of new energy electric vehicles, while large-scale access to the grid for charging will exacerbate the peak-valley difference and impact the security and stability of the power grid, and assess the stochasticity, uncertainty, and multi-timescale responsiveness, and correspond to the adoption of targeted optimization strategies to cope with the problem.

Glossary

Ancillary Services Grid-stabilizing functions provided by VPPs, including frequency regulation, voltage support, reserve capacity, and emergency response. VPPs earn compensation for these services, which are critical for maintaining grid reliability.

Big Data Platform A scalable infrastructure for processing, storing, and analyzing massive VPP data streams. It integrates tools for data cleaning, visualization, and predictive modeling to support informed decision-making.

Business Model The framework through which VPPs generate value, including energy trading, demand response, ancillary services, and capacity compensation. VPP business models integrate technical operations with market participation and financial incentives.

Capacity Compensation Financial incentives for VPPs that reduce peak grid demand or provide reserve capacity. This includes subsidies for load shifting, energy storage deployment, and grid stability support, enhancing VPP profitability.

Carbon Emission Trading The monetization of reduced carbon emissions through VPP operations, such as integrating renewables or optimizing energy use. VPPs can sell carbon credits in emissions markets, contributing to environmental sustainability and additional revenue.

Centralized Control Architecture A VPP management model where a single central control center directly oversees all resources, scheduling generation and consumption plans. Suitable for small to medium-scale VPPs, this architecture offers high control authority but faces scalability limitations due to high communication and computational loads.

Communication Architecture The network infrastructure and protocols used in VPPs to enable real-time data transmission between distributed resources, control centers, and the grid, supporting technologies like 5G, fiber optics, or power-line communication.

F. Yao, *Engineering Guide to Planning and Design of Virtual Power Plant*,
https://doi.org/10.1007/978-981-95-2231-6

Communication Network Layer The infrastructure layer responsible for bidirectional data transmission between VPP control centers, distributed resources, and grid systems. It utilizes technologies like 5G, fiber optics, and power-line communication (PLC) to ensure reliable, low-latency communication for real-time control and monitoring.

Controllable Loads Electrical loads that can be adjusted in real-time by a VPP, such as industrial machinery, commercial HVAC systems, electric vehicle charging stations, or residential appliances. Controllable loads enable VPPs to respond to grid demand, market prices, or dispatch signals, supporting peak shaving and demand-side management.

Cross-Regional Trading VPP participation in power markets across different geographical regions, enabled by flexible communication systems and standardized trading protocols. Cross-regional trading allows VPPs to leverage price differences, resource diversity, and regional grid needs.

Data Hardware Platform Layer The physical infrastructure (servers, edge devices, storage systems) that collects, processes, and stores operational data from VPP resources. It supports functions like data visualization, logging, and aggregation, enabling seamless integration of DER data for grid management.

Data Security and Privacy The protection of sensitive power operation data and consumer information within VPPs, addressing risks like data breaches or unauthorized access. Robust security measures are critical for maintaining trust and regulatory compliance.

Demand Side Management (DSM) Strategic management of electricity consumption at the customer side to optimize energy efficiency, often through government policies or commercial agreements to shift load from peak to off-peak periods.

Demand Side Response (DSR) Adjustments in electricity consumption by end-users in response to price signals, grid stress, or incentives, coordinated by VPPs. DSR enables load shifting, peak reduction, and revenue generation through participation in demand response markets.

Digital Technology Integration The use of advanced communication, computing, and control technologies (e.g., 5G, AI, blockchain) to enable VPP functionality, including resource aggregation, real-time monitoring, and optimized scheduling. Digital technologies are foundational to VPP operation and scalability.

Digital Twin A virtual, real-time model of physical VPP assets (e.g., power plants, microgrids, or grid infrastructure) used for simulation, optimization, and predictive maintenance. Digital twins enhance VPP operational efficiency and reliability.

Distributed Control Architecture A hierarchical model where local control centers manage regional resources, with a higher-level center coordinating overall operations. This approach reduces data congestion, improves scalability, and enables decentralized decision-making for larger VPPs.

Distributed Energy Resources (DERs) Small-scale energy generation, storage, or consumption systems distributed across the grid, including rooftop solar panels, small wind farms, battery storage, and flexible loads. VPPs integrate DERs to enhance grid flexibility, optimize energy use, and participate in market services.

Economic Analysis Evaluation of VPP project costs (e.g., capital expenditure, operational expenses) and revenue streams (e.g., energy sales, ancillary services) to assess financial viability and return on investment.

Edge Computing Distributed computing at the network edge to process real-time data from DERs, reducing latency and bandwidth requirements. Edge devices in VPPs enable local decision-making for faster grid response and resource optimization.

Electricity Market A competitive marketplace where electricity producers and users trade energy and related products, with prices and quantities determined by market competition. VPPs participate as unified entities in such markets.

Energy Storage Devices Systems that convert and store electrical energy for later use, including lithium-ion batteries, pumped hydro storage, and compressed air energy storage. Energy storage allows VPPs to mitigate renewable energy intermittency, provide grid ancillary services (e.g., frequency regulation), and optimize energy trading through peak-valley arbitrage.

Federated Learning A privacy-preserving AI technique where models are trained on decentralized data without centralizing sensitive information. This allows VPPs to collaborate on AI models while protecting consumer data privacy.

Financial Analysis The evaluation of VPP project costs (capital expenditure, operational expenses) and revenue streams (energy trading, subsidies, ancillary services). It involves modeling cash flows, risk assessment, and return on investment (ROI) to assess economic viability.

Full Distributed Control Architecture A decentralized model where autonomous VPP subsystems operate independently, with a central hub serving as a data exchange center. This architecture offers exceptional scalability and flexibility for large, cross-regional VPPs but may have slightly weaker real-time control over terminal resources.

Generative AI AI models that simulate extreme grid scenarios (e.g., cyberattacks, natural disasters) to enhance VPP resilience. Generative AI helps test contingency plans and optimize disaster response strategies.

Grid Scheduling The coordination of VPP resources to align with grid operator instructions, ensuring supply-demand balance, voltage stability, and compliance with operational constraints. Grid scheduling is essential for VPP integration into the power system.

Intelligent Application Layer The software layer that enables advanced VPP functions, including demand response, ancillary service participation, and market trading. It uses artificial intelligence (AI), machine learning, and big data analytics to optimize resource scheduling and revenue generation.

Intelligent Control System Advanced systems using IoT, big data, and AI to monitor, analyze, and optimize VPP operations, enabling autonomous decision-making for energy dispatch, market trading, and resource coordination.

Microgrid A localized energy network that can operate independently or in conjunction with the main grid, often centralized geographically. Unlike VPPs, microgrids manage their own physical infrastructure, while VPPs aggregate dispersed resources via communication networks.

Multi-Agent Reinforcement Learning (MARL) A machine learning approach where autonomous AI agents coordinate to optimize VPP operations. MARL enables self-sufficient grid management, allowing DERs to adapt dynamically to changing conditions without human intervention.

Multi-Ecosystem Integration The aggregation of diverse resources (e.g., electric vehicles, distributed energy storage, building loads) into a unified VPP network. This integration enables participation in grid services like peak shaving, frequency regulation, and cross-market trading.

Phased Construction Scale The staged planning and implementation of VPP projects, allowing for incremental expansion of distributed energy resources, hardware, and software platforms to adapt to evolving needs and technologies.

Pilot Sample Design A demonstration project designed to validate VPP functionality in real-world scenarios. It involves integrating DERs (e.g., solar, energy storage, controllable loads) and testing their coordination for grid stability, market participation, and economic optimization.

Power Market Transactions The buying and selling of electricity by VPPs in wholesale or retail markets, including spot trading, futures contracts, and ancillary service auctions. VPPs optimize transactions to maximize revenue and meet grid demand.

Protection System Design Measures to safeguard VPPs against faults and disruptions, including bidirectional power flow protection, phasor measurement units (PMUs), and AI-driven predictive maintenance to ensure grid stability.

Quantitative Risk Analysis The assessment of uncertainties in VPP operations, including market price volatility, policy changes, and technical failures. It uses statistical models to quantify risks (e.g., spot market revenue fluctuations) and develop mitigation strategies like diversification or hedging.

Renewable Energy Integration The process of incorporating variable renewable energy sources (e.g., solar, wind) into VPPs, leveraging energy storage and controllable loads to mitigate intermittency and enhance grid acceptance.

Response Time The duration required for VPPs to adjust power output or load in response to grid signals or market changes, categorized into day-ahead, intraday, and real-time schedules.

Scalable Resources Distributed energy resources in VPPs that can be easily expanded or modified, such as additional solar panels, energy storage units, or controllable loads, to accommodate growth.

Smart Sensing and Actuation The deployment of sensors and actuators in energy resources to enable intelligent monitoring and edge control. This technology allows real-time data collection on parameters like voltage, current, and equipment status, enabling dynamic adjustment of distributed energy resources (DERs) within a VPP.

Software-Network-Hardware Synergy The integration of software applications, communication networks, and hardware devices to enable seamless VPP operation. This synergy ensures efficient data flow, resource control, and market participation.

System Resilience The ability of VPPs to maintain functionality during grid disturbances (e.g., outages, extreme weather) by leveraging distributed resources, energy storage, and autonomous control, ensuring continuous operation and grid stability.

Technical Specifications Detailed requirements for VPP hardware and software, including generation capacity, regulation rate, communication latency, and measurement accuracy, to ensure grid compatibility and operational reliability.

Time-of-Use Tariffs Electricity pricing structures that vary by time of day, incentivizing VPPs to shift load consumption to off-peak hours for cost savings or revenue generation through energy arbitrage.

Time-Series Database A database optimized for storing and managing time-stamped energy data (e.g., power generation, load profiles). It supports real-time analytics and historical trend analysis for VPP performance evaluation.

Virtual Power Plant (VPP) An intelligent aggregation of distributed energy resources (DERs)—such as solar PV, wind turbines, energy storage systems, and controllable loads—managed via digital technology to function as a unified, controllable entity in power market transactions and grid operations. VPPs optimize resource coordination without requiring extensive hardware investments, enabling flexible participation in grid scheduling and energy trading.

Virtualization The digital representation of physical VPP components, enabling cross-regional resource aggregation and flexible management. Virtualization allows VPPs to operate as unified entities regardless of geographic distribution.

Index

F. Yao, *Engineering Guide to Planning and Design of Virtual Power Plant*,
https://doi.org/10.1007/978-981-95-2231-6

MIX
Papier aus verantwortungsvollen Quellen
Paper from responsible sources
FSC® C105338

If you have any concerns about our products,
you can contact us on
ProductSafety@springernature.com

In case Publisher is established outside the EU,
the EU authorized representative is:
Springer Nature Customer Service Center GmbH
Europaplatz 3, 69115 Heidelberg, Germany

Printed by Libri Plureos GmbH
in Hamburg, Germany